NIKOLA TESLA: LECTURE BEFORE THE NEW YORK ACADEMY OF SCIENCES

April 6, 1897

Nikola Tesla
International Telecommunications Union

NIKOLA TESLA: LECTURE BEFORE THE NEW YORK ACADEMY OF SCIENCES

The Streams of Lenard and Roentgen and Novel Apparatus for Their Production

April 6, 1897

Reconstructed

Leland I. Anderson, Editor

1994

TWENTY FIRST CENTURY BOOKS BRECKENRIDGE, COLORADO

Library of Congress Catalog Card Number: 94-61004

ISBN 0-9636012-1-0 (hard cover)
0-9636012-7-X (soft cover)

Twenty First Century Books
P.O. Box 2001
Breckenridge, Colorado 80424

Contents

Figures *vii*

Editorial Remarks *viii*

Preface *ix*

Introduction *xiii*

Background 1

- Setting 1
- Skirmishes on non-publication of lecture 3

Lecture Commentary 7

- High frequency apparatus 7
- Lenard and Roentgen rays 18
- Harmful actions from Lenard and Roentgen tubes 26

The Lecture:

- Section I - Improved Apparatus for the Production of Powerful Electrical Vibrations; Novel Frequency Measurement Methods. 29
- Section I Addendum - Wireless Telegraphy Receiving Methods. 71
- Section II - The Hurtful Actions of Lenard and Roentgen Tubes. 83
- Section III - The Source of Roentgen Rays and the Practical Construction and Safe Operation of Lenard Tubes. 95

Appendix 109

- Contemporary reviews of lecture 111

Acknowledgments 117

Sponsorship 118

Index 121

Figures

Form of listing: Sec. #.-Fig. #

LC	(Half-wave node in primary circuit)	14
I-1	Method of transformation of electrical energy by oscillatory condenser discharges	34
I-2	Mechanical analogy of electrical oscillator	35
I-3	System illustrated in figure 1 with self-induction coil	37
I-4	Coil wound to secure greatly increased capacity	37
I-5	Associating a secondary coil with a primary circuit coil	38
I-6	System adopted for existing municipal circuits	38
I-7	Circuit controller allowing condensers connected to discharge circuit alternately and successively	39
I-8	Arrangement of parts and circuits of a small oscillator	39
I-9	Photograph of small oscillator diagrammatically shown in figure 8	41
I-10	Apparatus for the manufacture of condensers and coils	46
I-11	High potential coil system having terminals at centers	48
I-12	Photograph of coil system illustrated in figure 11 in action	52
I-13	Special instrument to exactly determine wavelength and phase	62
I-14	Method of impulse illumination of instrument disk	67
IA-15	Devices for receiving	72
IA-16	Other ways of receiving	73
IA-17	(Segmented-armature interrupter)	74
IA-18	(Mercury-break interrupter)	74
IA-	(A series of six photographs of drawings of 120 bulbs exhibited on walls of New York Academy of Sciences)	76-81
III-1	Illustrating an experiment revealing the real source of the Roentgen rays	97
III-2	Improved Lenard tube	102
III-3	Illustrating arrangement with improved double-focus tube for reducing injurious actions	103
III-4	Illustrating arrangement with a Lenard tube for safe working at close range	106

Editorial Remarks

Section I of this lecture is presented with few changes from the original text prepared by Tesla, an illustration of which is reproduced on page 30. The text would have benefited from an editor's hand if presented to a publisher at that time, but no such editorial "smoothing" has been attempted in its presentation now. Only minor changes have been introduced, such as in words that were separated before the turn of the century but now appear solid. They are: electro-magnetic, electro-motive, in as much, tin foil, and wave length. A few articles and prepositions were missing, and their obvious replacements have been set in brackets [ ]. As an additional aid to the reader, certain items have been marked in brackets with an explanation provided in a note. Line-drawing figures 13 ab, 14, and 16, together with the series of six photographs of drawings of 120 bulbs appearing on pages 76-81, have been enhanced.

Section I appeared among a group of papers passed on to his children by George Scherff, who was Tesla's personal secretary, business manager, and confidant from 1895 through 1939. A copy also exists in the archives of the Tesla Museum in Belgrade. Both copies, however, were missing some illustration drawings and photographs. These were drawn from the archives of Knight Brothers and Boyle & Anderson.

Preface

Nikola Tesla was born of Serbian parents at Smiljan, in the Austro-Hungarian border province of Lika, now part of Croatia, at midnight July 9-10, 1856. His father, Milutin, was a Serbian Orthodox priest, and his mother, née Djouka Mandić, was of a family line whose sons were of the clergy and whose daughters were wives of the clergy. The Serbian Orthodox church then used the Julian calendar, and it continues to use this calendar today for days of observance. The American colonies had converted to the Gregorian calendar 132 years before Tesla arrived at New York in 1884. When Tesla crossed this 'date line,' 11 days dropped from his personal calendar. Most institutions observe Tesla's birth date as July 10, which date Tesla held for himself, but if the tolling church bells in Lika could have been heard in America when Tesla was born, the calendar date would have been July 21, 1856.*

Establishing himself in the United States, Tesla became a citizen in 1891. He brought to the world great inventive gifts for which he became famous—the induction motor and the multi-phase alternating current power distribution system driving it (1888); the fundamental system of wireless telegraphy embodying the "Tesla coil" (1893); telemechanics (1898); the Tesla turbine (1913-20), which is attracting great revival interest; and, among many leading inventive achievements, VTOL aircraft (1928).

* The Tesla family moved to nearby Gospić when Nikola reached the age of six to enter school. Adding to the uncertainty of the date accorded Tesla's birth is an official certificate of birth entered for him by the city of Gospić showing his birth date as June 28, 1856. This certificate is reproduced in *Nikola Tesla: Correspondence with Relatives* (Belgrade: Nikola Tesla Museum, 1993; in Serbian and English), illus. sec. I thank Milan Radović, University of Wisconsin-Madison Libraries, for translating this significant document.

For 50 years following the presentation of the principles of wireless telegraphy now called radio at his lecture in St. Louis, Tesla asserted his inventive claim. It wasn't until five months following his death in 1943 that the United States Supreme court declared the basic radio patent of Marconi invalid, recognizing the prior art of Tesla for the system concept and apparatus, Stone for the method of selectivity, and Lodge for variable tuning.

In his lifetime, Tesla was granted over 30 honorary degrees and foundation medals from the world over. The unit of magnetic flux density in the MKS system was named "tesla" on the occasion of the centennial year of his birth. The only other American to share such recognition is Joseph Henry.

The 1897 lecture before the New York Academy of Sciences was the sixth of historic lectures delivered in rapid succession in America and in Europe. The previous five lectures were:

> A New System of Alternate-Current Motors," May 16, 1888, before a meeting of the American Institute of Electrical Engineers (AIEE) in New York City; followed by the *trio series* of demonstration lectures on high frequency and high potential alternating currents, the first,"Experiments with Alternate Current of Very High Frequency and Their Application to Methods of Artificial Illumination," May 20, 1891, before a meeting of the AIEE in New York City; the second, "Experiments with Alternate Currents of High Potential and High Frequency," February 3, 1892, before the Institution of Electrical Engineers, London, followed a day later (with some condensation) by special request at the Royal Institution, and by invitation, February 19, before the Société Internationale des Electriciens and the Société Francaise de Physique; and the third, "On Light and Other High Frequency Phenomena," February 24, 1893, before the Franklin Institute in Philadelphia and (with some variation) March 1, before a meeting of the National Electric Light Association in St. Louis (it was in the latter that the principles of radio telegraphic communication were first presented); and "Mechanical and Electrical Oscillators," August 25, 1893, before a meeting of the

International Electrical Congress at the Columbian Exposition in Chicago, and (with some variation) November 29, before a meeting of the New York Electrical Society.

Tesla delivered four additional technical lectures or papers *in absentia*, the last in 1911.

Introduction

There are more things in heaven and earth, Horatio,
than are dreamt of in your philosophy.

Hamlet, Act 1, Scene 5

In 1895, the fluorescent coating of a cardboard screen offered just enough illumination for Professor Wilhelm Conrad Roentgen to find his way to the discovery of the X ray. The implications of his discovery for the medical profession were, of course, no less than staggering. Yet history has shown that its implications for the world of physics were more far-reaching than anyone could have imagined.

At the time of Roentgen's discovery, many scientists were quite comfortable with the Newtonian explanation of the way the universe worked and even discouraged students from pursuing careers in physics because, as they thought, physics offered little career potential: almost everything already had been explained! A few notable exceptions, among them black-body radiation and the Michelson-Morley experiment, challenged the neat and simple explanations of Newtonian physics, but the world of science was confident that these phenomena soon would be understood. The nature of light and electricity (ether waves vs. radiant matter) and the structure of the atom were still in the question-forming stage. Roentgen's discovery of the X ray signaled the end of two centuries' confidence in Newtonian physics. With little reservation, the discovery of the X ray can be considered the birth of modern physics.

Where, boundless nature, can I hold you fast?

Faust, 455

The trail leading to the discovery of Roentgen's invisible light began in the seventeenth century with two chance observations of visible light: Von Guericke noting that a faint glow occurred between his hand and a spinning sulfur ball and Piccard finding that light is emitted from mercury sloshing around the top of a barometer. By the early eighteenth century, Hawksbee, knowing that a vacuum had been present in the upper part of Piccard's barometer, constructed glass vessels removing some of the air. He excited them with frictional electricity and observed beautiful glowing streams of colored light. Almost a century and a half passed before the real significance of this effect—the accelerated electron—would begin to be appreciated.

The 1790 introduction of Volta's electric battery allowed Oersted, Ampère, Faraday, and Henry to deduce the relationship between electricity and magnetism. Page and Rhumkorff constructed induction coils, the first high voltage transformers, which could increase the few volts from a battery to many thousands of volts. In the 1850s, Geissler developed an efficient vacuum pump, similar to Piccard's barometer, which employed the weight of mercury to pull the air out of a glass vessel. Geissler, a glass blower, fabricated many types of glass tubes, evacuated them using his own new pump, and watched them glow with beautiful colors when high voltage from induction coils was applied. Variations in the glass composition, the kind of gas, and the level of vacuum were used to expand the multicolor effects of Geissler tubes.

As equipment improved, higher vacuums were attained; Plücker, Hittorff, and Crookes observed streams of light emanating from the negative electrode of some of their discharge tubes. Furthermore, a magnet was seen to bend and deflect this stream, called the cathode ray. Crookes, Goldstein, and Perrin designed many variations of vacuum discharge tubes, most of which demonstrated new facets of the cathode ray. Magnetic-effect, paddle-wheel, and canal-ray tubes were only a few of such types.

In the nineteenth century, the true nature of cathode rays was a subject of much controversy. In 1803, Thomas Young showed that light had a wave nature and many scientists assumed that cathodic light (one name given to cathode rays) was just another light wave traveling in the ether. However, Crookes, among others, saw the cathode ray as a stream of matter particles which he termed "radiant matter." In 1890, Hertz constructed an experiment in which cathode rays were found to exit a glass discharge tube through a thin aluminum window opposite the cathode. His untimely death in 1894 left his student, Lenard, with the task of continuing their experiments. Lenard asked Roentgen to repeat some of the experiments that Hertz and he had conducted and presented Roentgen with some apparatus with which to accomplish them.

Like many other scientists of his day, Roentgen focused on the cathode ray. On the evening of November 8, 1895, he carefully covered a discharge tube with a black cardboard to prevent the light in the tube from interfering with his investigation. Immediately upon energizing the tube, Roentgen noticed a greenish glow emanating from a nearby cardboard screen that had been coated with a chemical compound known to fluoresce in the presence of cathode rays. Cathode rays had never been known to journey more than a few centimeters from their discharge tube, and since the cardboard screen was more than a meter's distance from the tube, Roentgen concluded that the glow he observed was the effect of a new kind of ray. During the course of further investigation, Roentgen chanced to expose his fingers to the ray as he held small lead fishing weights in front of the discharge tube: The shadow of the bones in his fingers were cast on the fluorescent cardboard.

The great discovery was made. Given the certainty that X rays were being produced in copious quantities by many of the discharge tubes in use at the time, X rays might have been discovered by any number of scientists during the previous several decades. Rather than diminishing Roentgen's achievement, however, this fact magnifies it considerably, demonstrating not only Roentgen's genius in recognizing what was happening but his wisdom in stopping to investigate. On December 28, 1895, Roentgen published about 50 copies of his preliminary paper and distributed them to his local colleagues in Bavaria. Early in 1896, he made the announcement to the rest of the world.

The idea of seeing the living inner bones and tissues of their bodies was disconcerting to most people, to say the least.

A New Jersey legislator proposed a bill to outlaw the making of X-ray opera glasses, while a London manufacturer offered X-ray-proof undergarments. In Roentgen's own culture, the sight of bones presaged imminent death, and Roentgen's wife was horrified by the image in her hand.

Obviously, the most immediate application of the new discovery was in the medical world, and medical practitioners, scientists, and instrument companies craved any information they could get. Crookes-type vacuum discharge tubes and induction coils were not easy to find outside of university physics laboratories. A Boston dentist, William J. Morton, actually made use of a simple light bulb connected to a borrowed induction coil to produce some of the first X-ray images in the United States. On March 11, 1896, the New York *Electrical Review* answered the clamor for information about X rays by launching a series of eight articles by Nikola Tesla in which he presented many new ideas, inventions, and research dealing with the X ray, its production, use, and scientific explanation.

The Röntgen Rays, The Röntgen Rays,
What is the craze,
The town's ablaze
With the new phase
of X-Ray ways . . . .

Wilhelmina
Electrical Review (London)
April 17, 1896

Anderson's reconstruction of Tesla's lecture before the New York Academy of Sciences on April 6, 1897, is a most important contribution. In this lecture, Tesla went beyond his titled topic, "The Streams of Lenard and Roentgen and Novel Apparatus for Their Production," and expanded on his X-ray articles published in the New York *Electrical Review*. Some 120 large drawings of his vacuum tubes were displayed on the walls of the lecture hall. Among the tubes represented were not only Crookes and Lenard types but varieties of single-electrode tubes of Tesla's own invention, some of which were used for his Roentgen-ray demonstrations accompanying the lecture.

During the lecture, Tesla discussed the uses of some of the tubes in his experiments with wireless telegraphy. Among his tubes, Tesla said, were "a great number of receiving devices." Compare this statement with Tesla's 1916 remark reported by Anderson in the first book of this series, *Nikola Tesla On His Work With Alternating Currents*: "Well, in some of these bulbs I have shown, for instance, that a heated conductor emits a stream of ions, or as I said at that time, charged particles, and a few of these bulbs have been used exactly in the same manner the audion is used today."

One is prompted to ask, "What was Tesla really looking for in his research and experimentation with vacuum discharge tubes?" His statements about using the tubes in the receiving and detection of wireless signals offers clues. In his lecture before the Academy, Tesla often referred to Lenard- and Roentgen-streams and tubes; he obviously considered Lenard and Roentgen to hold equal priority in the discovery of the X ray. The Lenard tube, as well as Lenard's experimentation, were of particular interest to Tesla in that cathode rays (streams of electrons) actually emanated from the aluminum window opposite the cathode of the tube and proceeded a few centimeters into the air. Recent research by James Corum and Kenneth Corum indicates that Tesla was looking for methods of moving electrons with such devices as open-air diodes or even relativistic electron-beam (REB) diodes, which, if viewed as they are built today, including power supply, resemble closely a Tesla coil and a Lenard tube. The speculation about Tesla's moving electrons is, perhaps, only the beginning of the story. Other evidence, particularly his research on particle beam weaponry, points to his interest in moving larger particles.

Another great value of Anderson's contribution in reconstructing this lecture is that it shows us historically the extent of Tesla's work with vacuum tubes up to 1897. Roentgen's announcing the discovery of the X ray provided Tesla with yet another area in which to contribute his discoveries and inventions. This lecture on the rays of Lenard and Roentgen, and Tesla's series of articles in the New York *Electrical Review* contain material far more advanced than any other contemporary work. Tesla's 1897 lecture discussion of "reflected" Roentgen rays, offered with data and tables, compares almost exactly to Arthur H. Compton's 1922 monograph on the topic of secondary radiation. Tesla also offered a design of apparatus to generate "reflected" rays. Tesla's lecture is a fount of information beyond the knowledge of most of his contemporaries, offering a wide array of tables, charts, diagrams, photographs, schematic designs, and suggestions of one process after another for the production of X rays, the use of vacuum tubes, and special procedures for refining the operation of all kinds of apparatus. Tesla, here and in other works, discussed scientific principles not "discovered" until years later.

The most significant contribution of this text is, perhaps, that it shows in his true light one of the greatest visionaries that ever lived—a man far ahead of his peers yet gentle and willing to give what he had to the world. With regard to the relationship between Tesla's work and the world of science, it is curious to note that it has taken a better part of the last 100 years since his invention of the resonant transformer for researchers to truly succeed in duplicating the Tesla coil design, this in spite of the great pains he went to in making his recommendations clear. Let us hope that, as more information on Tesla's work becomes available, much greater attention will be given to it, to the betterment of our world.

Jim Hardesty
Judith Hardesty
Ithaca, New York
June 1994

One cannot help looking at that little bulb of Crookes with a feeling akin to awe, when he considers all that it has done for scientific progress—first, the magnificent wonderful achievements of Roentgen. Possibly it may still contain a grateful Asmodeus, who will be let out of his narrow prison cell by a lucky student. At times it has seemed to me as though I myself heard a whispering voice, and I have searched eagerly among my dusty bulbs and bottles. I fear my imagination has deceived me, but there they are still, my dusty bulbs, and I am still listening hopefully.

N. Tesla
March 7, 1896

Background

Setting

The 1897 lecture before the New York Academy of Sciences did not appear in its entirety in Tesla's lifetime. In an extended 1916 interview he remarks,

> "The lecture was not published because I had to do a lot of things. I had undertaken an extensive program, and I found that my energies were not adequate to the task. Later on, the subject was neglected; other business prevented me from doing anything else. It only exists in typewritten form, uncompleted."[1]

The original 1897 lecture, as delivered, carried the title, "The Streams of Lenard and Roentgen and Novel Apparatus for Their Production,"[2] but in actuality it went far beyond that topic. On the walls of the lecture hall Tesla displayed approximately 120 large drawings of vacuum tubes that he ordered built in several forms by his laboratory technicians. Many of these were of the Lenard type and also the single-electrode type pioneered by him and used in demonstrations of X-ray[3] methods in the lecture. Among the drawings were tubes used in wireless telegraphy experiments. The heretofore unpublished first portion of the 1897 lecture text covers, with the exception of Tesla's opening remarks on X-ray

[1] Anderson, L., ed., *Nikola Tesla On His Work With Alternating Currents and Their Application to Wireless Telegraphy, Telephony, and Transmission of Power* (Denver: Sun Publishing, 1992), p. 158.

[2] (Editorial) "The New York Academy of Science: An Interesting Electrical Exhibition—Address by Nikola Tesla announcing recent achievements," *Electrical Review* (N.Y.), Apr. 14, 1897, p. 175; (Editorial) "Mr. Tesla on Roentgen Rays," p. 398, and (Review) "Mr. Tesla Before the New York Academy of Sciences," *Electrical Engineer*, Apr. 14, 1897, pp. 400-401; the latter was published under the title, "Mr. Tesla on X Rays," in the *Electrical Review* (London) May 7, 1897, p. 626. See Appendix for reproductions of these reviews.

[3] The term "X ray" had not, at the time of this lecture, been generally accepted for the rays of Roentgen.

discovery, high frequency resonators and measurement methods. In addition, Tesla covered an extension of the measurement topic into wireless telegraphy receiving methods which is presented as an Addendum. These topics are not suggested by the lecture title.

It is concluded that the subject matter of the lecture specifically covered by its title was published over Tesla's name in communications entitled "Tesla on the Hurtful Actions of Lenard and Roentgen Tubes" and "Tesla on the Source of Roentgen Rays and the Practical Construction and Safe Operation of Lenard Tubes," *Electrical Review* (N.Y.), May 5 and August 11, 1897, respectively. These portions of the lecture immediately follow the Addendum. With exception of the first five introductory sentences of the second communication referencing the lecture before the Academy, the three segments plus added remarks from the 1916 interview (i.e., the Addendum) allow one continuous reading as the complete lecture.

The outline of the reconstructed lecture is as follows:

Section I

Improved Apparatus for the Production of Powerful Electrical Vibrations; Novel High Frequency Measurement Methods.

Section I - Addendum

Wireless Telegraphy Receiving Methods.

Section II

The Hurtful Actions of Lenard and Roentgen Tubes.

Section III

The Source of Roentgen Rays and the Practical Construction and Safe Operation of Lenard Tubes.

There are perhaps two combining reasons why the lecture was not published by Tesla in its entirety as delivered. The first, as he points out in the opening remarks of the lecture and in the 1916 interview, relates to the intense research and engineering activity he was at that time engaged. Some of these efforts include (a) the crystallization of his research work in wireless telegraphy, telephony, and transmission of power in the form of patent coverage; (b) the work on wireless-controlled *telautomata*; (c) the development of high intensity phosphor-coated vacuum-tube lamps—the fluorescent lamps of today; and (d) the time-consuming task defending his patent rights in the courts.

The second reason relates to the growing, highly competitive atmosphere in wireless telegraphy development now recognized as radio. Preparing at that time applications for the fundamental patents in wireless telegraphy,[4] he undoubtedly realized that the detailed information contained in this portion of the lecture (presented as Sect. I) was of proprietary value and it would be given away without first obtaining protection through patent applications. Whereas Tesla had been giving freely to the world many technological ideas in the course of his classical *trio series* of lectures delivered in 1891, '92, and '93,[5] he now grew more cautious, and rightfully so, in protecting his own economic interests in the field of wireless communication.

Skirmishes on non-publication of lecture

At that time in the United States, six weekly commercial electrical journals were published: *Electrical Engineer*, *Electrical Review*, *Electrical World*, *Electrician*, *Electricity*, and *Western Electrician*. In addition to these were the weekly *Scientific American*, which reported on a mix of scientific and engineering topics, as well as the monthly engineering periodicals *Electrical Engineering* and *Engineering*

[4] U.S. Patent No. 645,576 of March 20, 1900, "System of Transmission of Electrical Energy," and U.S. Patent No. 649,621 of May 15, 1900, "Apparatus for Transmission of Electrical Energy," both applications filed Sept. 2, 1897.

[5] Martin, T.C., *The Inventions, Researches and Writings of Nikola Tesla* (New York: The Electrical Engineer, 1893), chs. 26-28 (pp. 145-373).

Magazine. This early publication activity was primarily a result of the vast electrical industrialization in America and new business opportunities that sprung from the inventions of Edison in the direct-current realm and, subsequently, the inventions of Tesla in the alternating-current realm. The marketplace was unquestionably saturated in electrical news coverage, and commercial electrical journals fought to stay alive.

Thomas C. Martin, serving as president of the American Institute of Electrical Engineers for the 1887-88 term, undertook editorship of *Electrical World* in December 1888 but was discharged in March 1890 over a bitter dispute with the owner of the publication on compensation. The feud was aired on seven tabloid-size pages in the September 30, 1893 issue.[6] Martin subsequently found a haven as editor of *Electrical Engineer*.

In 1893, the Electrical Engineer Company published the book, *The Inventions, Researches and Writings of Nikola Tesla*, compiled by Martin, its chief editor. This book remains a classic a century later having been republished several times.[7]

Seven years before his death Martin remarked in a letter to Elihu Thomson, "I made some money out of my Tesla book, which was promptly borrowed from me by the titular component, so that two years of work went for nothing."[8] The Electrical Engineer Company was undoubtedly in precarious financial condition as the letter suggests stock was offered to Martin in lieu of cash receipts from sales of the book.

[6] I am indebted to Marc Seifer, through his forthcoming work, *The Tesla Trilogy*, for this revealing insight concerning Martin's business relationship with Tesla.

[7] The first reprint was undertaken in 1952 by the Lee Engineering Co. of Madison, Wisc., as a contribution to the engineering industry.

[8] Jan. 16, 1917, in *Selections from the Scientific Correspondence of Elihu Thomson*, Abrahams, H.J., and Savin, M.B., eds. (Cambridge, Mass.: MIT Press, 1971).

During the year prior to the 1897 Academy of Sciences lecture, the *Electrical Review* published a remarkable series of communications from Tesla in the journal on the subject of material stream emanations.[9] Not to be outdone, Martin presumably persuaded Tesla to have the Academy lecture published in the *Electrical Engineer.* But Tesla obviously had second thoughts for reasons expressed earlier. Martin, perhaps not fully appreciating the role of an independent inventor depending for his livelihood upon the successful marketing of patented inventions, could not accept Tesla's decision to not complete the lecture for publication.

Martin had a ravenous appetite for material to make the *Electrical Engineer* financially successful. The following year, Martin allowed to be published in the November 17, 1898 issue, without permission, Tesla's paper, "High Frequency Oscillators for Electro-therapeutic and Other Purposes," presented before the eighth annual meeting of The American Electro-Therapeutic Association in Buffalo, New York, September 14, 1898. An editorial salvo entitled "Mr. Tesla and the Czar" also appeared in the same issue—a sort of 'tit for tat' rejoinder on the non-publication of the Academy of Sciences lecture—together with a specific patronizing editorial, "Tesla's Electrical Control of Moving Vessels or Vehicles from a Distance," dwelling on the impracticality of his recent invention.[10]

Martin certainly went beyond professional ethics in publishing the lecture. Undoubtedly, the stress of the financial problems of the *Electrical Engineer* was too trying. He needed 'leading-edge' material, and Tesla was not producing it.

[9] "Tesla's Startling Results in Radiography at Great Distances through Considerable Thickness of Substances," Mar. 11, pp. 48, 49; "Tesla's Latest Results," Mar. 18, pp. 146, 147; "Tesla on Reflected Roentgen Rays," Apr. 1, pp. 171-174; "Roentgen Radiations," Apr. 8, pp. 183-186; "Tesla's Latest Roentgen Investigations," Apr. 22, pp. 206-211; "Tesla Describes an Interesting Feature of the X-Ray Radiation," July 8, p. 43; "Roentgen Rays or Streams," Aug. 12, pp. 78, 79, 83; and "Roentgen Streams," Dec. 2, pp. 277, X.

[10] U.S. Patent No. 623,809 of Nov. 8, 1898, "Method and Apparatus for Controlling Mechanism of Moving Vessel or Vehicles," application filed July 1, 1898.

Tesla was furious concerning this unauthorized publication prior to the opportunity given the American Electro-Therapeutic Association to publish in its *Transactions,*[11] and sent copies of his severe objection to all electrical journals in America and England. The *Electrical Engineer* also carried the letter with accompanying defensive commentary running four columns in the November 24 issue.

To understand the financial crisis facing the *Electrical Engineer*, one must go ahead another three months to March 11, 1899, when the *Electrical Engineer* was taken over by the *Electrical World*, the merged publication becoming the *Electrical World and Engineer.* Martin was retained, but as co-editor. Given Martin's temperament, he perhaps did not perceive the move humbling as did others. Years later, in the fiftieth anniversary volume of *Electrical Engineering*, the official journal of the AIEE, Martin's misadventure with the *Electrical Engineer* was not mentioned.

It should not be assumed that Martin held a permanent grudge toward Tesla, even though later promoting Marconi in articles. Martin was too much of a gentleman to let such professional competition interfere with true friendship. But, the wound of the failed *Electrical Engineer* required some years to heal.

It was the endearing prose that he reserved for Tesla in his earlier articles appearing in the *Century Magazine,*[12] and these were not forgotten in 1911 when introducing him for an address on his new inventions before a meeting of the New York Section of the National Electric Light Association May 15, 1911. Martin reflected on Tesla's important lecture 18 years earlier before the Association, "On Light and Other High Frequency Phenomena," in which he presented the fundamental concepts of radio communication. And in the book which Martin co-authored with Stephen L. Coles, *The Story of Electricity* (1919), he remarked, "Tesla's influence may truly be said to have marked an epoch in the progress of electrical science."

[11] Tesla, "High Frequency Oscillators for Electro-therapeutic and Other Purposes," American Electro-Therapeutic Association *Transactions,* 1899, pp. 9-29. This paper has been reproduced by Twenty First Century Books, Breckenridge, Colo. (1994).

[12] "Nikola Tesla," Feb. 1894, pp. 582-585, and "Tesla's Oscillator and Other Inventions," Apr. 1895, pp. 916-933.

Lecture Commentary

High frequency apparatus

On March 13, 1895, Tesla's laboratory at 35 South Fifth Avenue (now West Broadway) in New York City was completely destroyed by fire—all the experimental and lecture apparatus, photographs, and documentation. Charles A. Dana editorialized,[13]

> *The destruction of Nikola Tesla's workshop, with its wonderful contents, is something more than a private calamity. It is a misfortune to the whole world. It is not in any degree an exaggeration to say that the men living at this time who are more important to the human race than this young gentleman can be counted on the fingers of one hand; perhaps on the thumb of one hand.*

On December 28 of this same year, Wilhelm Roentgen (1845-1923) presented his communication, "On a New Kind of Rays," to the Würzburg Physico-Medical Society, the news of it reaching American newspapers 11 days later. It was a discovery of universal significance, the ability to "see" images through otherwise opaque bodies, for which he was awarded the first Nobel Prize in physics in 1901. In Section I of the lecture that follows, Tesla describes in detail his researches along the same lines that undoubtedly would have led him to the same discovery were it not for the circumstances of the destruction of his laboratory by fire.

Edward R. Hewitt, grandson of Peter Cooper—inventor-philanthropist who founded Cooper Union in 1859—has given an interesting account of Tesla's lost opportunity.[14] During his experimental researches in photography, Hewitt

[13] *The Sun* (New York), Mar. 14, 1895, p. 6. c. 5.

[14] Busch, N.F., "Trout Fisherman—Sportsman of the old school, R. Hewitt calls himself 'last of the gentlemen mechanics,'" *Life*, July 15, 1946, pp. 86-100. See also reference to Hewitt's working with Tesla on X-ray photography in Thompson, E.P., *Roentgen Rays* (N.Y.: D. Van Nostrand, N.Y., 1896) p. 157. Hewitt is mentioned by Tesla in his April 22, 1896 communication to the *Electrical Review*.

describes a morning in Tesla's laboratory before the fire when attempting to photograph Mark Twain under a Geissler tube.[15] The resulting print was instead of the camera lens adjusting screw! Hewitt notes that neither he nor Tesla realized, until weeks later when Roentgen made his announcement, that this was an X-ray photographic image.

As the title of the lecture suggests, Tesla presented and described many pieces of apparatus for generating high frequency, high voltage alternating currents to excite Lenard and Roentgen tubes. Beginning with a discussion of the evolution the resonant-coil system named after him,[16] diagrammatically shown in figure 1, Tesla uses a fluid-mechanics analogue to illustrate the operation of charging and spark-gap components of the primary coil circuit associated with the condenser illustrated in figure 2. Tesla often employed analogic models to describe operating principles of the systems and apparatus of his invention. Whereas there are certain limits of applicability of such models for electromagnetic phenomena, views on this topic are not in general agreement.[17]

Beginning with figure 3 and extending through figure 8 of Section I, Tesla shows his innate understanding of the practical operation of these resonant-coil systems to produce powerful, high frequency and high potential alternating

[15] For a full-page photograph of Mark Twain with Tesla in his laboratory, see Martin, T.C., "Tesla's Oscillator and Other Inventions," *Century Magazine*, Apr. 1895, pp. 916-933 (photo appears on p. 930).

[16] U.S. Patent No. 462,418 of Nov. 3, 1891, "Method and Apparatus for Electric Conversion and Distribution," application filed Feb. 4, 1891.

[17] See Trifunović, Dragan, "Jedno uopštenje zakona o centralnim silama - o odnosu Nikole Tesla i Đorđem Stanojevića" (A Generalization on the Law of Central Forces - About the Relation Between Nikola Tesla and Djordje Stanojevich), in Transactions, Simpozij "Nikola Tesla," Zagreb–Smiljan/Gospić, July 7-10, 1976, pp. 203-211. In this paper, Trifunović discusses certain analogy problems in science and refers to the correspondence between Tesla and physicist Stanojevich during the time he was engaged in the discovery of analogies between the disparate phenomena of gravitational, magnetic, electrical, electromagnetic, optical, botanic- and biologic-cell, and planetary fields.

currents. The efficient design of physically small-size coils operating from existing municipal circuits is discussed, especially those for use in physicians' offices then supplied with 110 volts direct current or 60/125 cycles per second[18] alternating current. Such coils were widely used by physicians in the early 1900s for electrotherapeutic purposes.[19]

Proceeding to the diagram shown in figure 7, Tesla has introduced a mechanical controller in the primary circuit allowing the alternate charging and discharging of independent condensers in the primary circuit from the power source. He also refers to a modification of this circuit having "one continuous contact common to the two circuits, and independent interrupters for each of these," allowing not only an alternate charging and discharge of the condensers but their simultaneous charging and discharge in parallel. These circuits are embodied in Tesla's United States patent No. 577,670 of February 23, 1897 (application filed Sept. 3, 1896). The importance of this technique was obviously not recognized by those attending the lecture—it was the heart of Tesla's later work on what he referred to as the "art of individualization" and embodied in his invention known today as the AND logic gate.

Coming to the diagram of figure 8, the layout of circuit elements is given for the desk-top coil unit shown in figure 9, which Tesla offered for use in operating X-ray tubes and various laboratory appliances. The unit stood 18 inches high at the discharge terminals and comprised several novel features giving an energy conversion efficiency of 80 percent. Four months later, when Lord Kelvin attended a meeting of the British Association for the Advancement of Science in

[18] Although the unit "Hz" for frequency is now universally adopted for frequency measurement, at the time of publication of this 1897 lecture, and for some eight decades later, "cycles per second" was the unit of measurement. To avoid reading inconvenience using Hz in editorial discussion and cycles per second in the lecture text, cycles per second (cps) will be used throughout this work.

[19] See, for example, Monell, S.H., *High Frequency Electric Currents in Medicine and Dentistry* (N.Y.: William R. Jenkins Co., 1910) and Eberhart, N.M., *A Working Manual of High Frequency Currents* (Chicago: New Medicine Publishing Co., appearing in eight editions from 1911 to 1923).

Science in Toronto, Tesla presented one of these units to him on the occasion of his visiting Tesla's Houston Street laboratory in New York City. Tesla produced several of these units, and later proceeded to enter into a business venture with a Mr. Hopkinson for the manufacture of a few sizes of coils and X-ray units, but information about the establishment of this business enterprise has not been located.[20]

Tesla next briefly describes his work with phosphor-coated bulbs to replace the incandescent-filament bulbs developed by Edison. He became interested in developing a lamp that would equal the intensity of sunlight, and in January, 1894, the first photograph taken in Tesla's laboratory by light only from his phosphorescent lamp appeared in the April, 1895 issue of the *Century Magazine* . It was an eight-minute exposure, but in a little more than two years Tesla had achieved such brilliance in his lamps that the May 20, 1896 issue of the *Electrical Review* (N.Y.) carried an illustration of only a two-second photograph of Tesla taken with a lamp of 250 candlepower—again by the light of the bulb itself. Such a result with the combination of the best existing film emulsions and indoor lighting had not before been achieved.

This single-electrode lamp, first shown by Tesla in his 1891 lecture—the first of his *trio series* of lectures delivered in America and Europe during the years 1891-93—was patented by him that year.[21] Following this public demonstration, Elihu Thomson filed a "System of Electrical Lighting" patent application that was determined to be in interference with Tesla's patent. Tesla's laboratory assistants testified that he did actually produce operating lamps with a single conductor and lead-in wire. However, Thomson was unable to prove such a demonstration, although asserting earlier work on the general subject, and his claim was therefore denied—the Patent Office declaring Tesla's priority in this invention.[22] In an unpublished statement entitled "Tesla's Artificial Daylight," written by Tesla and designed to secure

[20] Tesla letter to George Scherff, Oct. 13, 1905 (Columbia University Libraries).

[21] U.S. Patent No. 454,622 of June 23, 1891, "System of Electric Lighting," application filed Apr. 25, 1891.

[22] U.S. Patent Office Interference No. 17334, *Thomson vs. Tesla*, notice filed July 12, 1895, decision of priority to Tesla June 10, 1897.

investment in a company to advance the application and manufacture of these lamps, the merits of this form of illumination were described:

> "The present systems [interior incandescent lamps and street arc lamps] are not only wasteful, they are also inseparable from other disadvantages as, the large cost of installation, which is chiefly due to the great quantity of copper required; the frequent renewal of the lamps, owing to their unavoidable deterioration; the disagreeable character of the light which, coming from a small surface, is naturally too intense and detrimental to the eyesight; the necessity of employing more or less opaque screens, which involve a considerable loss in illuminating power, and many other drawbacks of this nature. It is true, that recently Nernst and others have secured some gain in the efficiency of incandescent lamps or burners by the use of coatings of rare oxides, which permit higher degrees of incandescence. But this departure has not done away with the objectionable features above mentioned—on the contrary, it has only added to them.
>
> In Tesla's new lighting system, all these disadvantages are successfully removed. The light is produced with a much smaller expenditure of energy, scarcely more than one eighth of what is presently consumed for the same quantity of light, it is soft and agreeable to the eye, closely resembling daylight... Tesla's system offers features of an ideal character. The lamps may be of any desired candlepower and may be adapted to any kind of current of supply, and they last indefinitely."

We find that Tesla, one of the early energy conservationists in engineering, was describing a number of the lighting apparata found in use today, prominently fluorescent-lamp systems.

All of Tesla's lectures from 1891 dwelt at length on the necessity of completely avoiding air bubbles in the condenser and coil units of high frequency, high voltage disruptive-discharge oscillators. Not only amateur experimenters and researchers, but professional engineers and physicists have overlooked the harmful and destructive effects of allowing air bubbles in such reactive components.

This editor has witnessed the destruction of a costly high-power, high-voltage transformer used in a particle accelerator laboratory because the power was initially applied too quickly, rather than raising the primary voltage gradually, over days, allowing the coils to "cook" air bubbles out of the oil bath. As well, most amateur experimenters and researchers have not taken the diligent, perfection care in insulating the components of their systems to the degree that Tesla advocated. It is often expressed, "How did Tesla achieve those brilliant results?—I seem not to be able to reproduce them with the same effectiveness as he claimed."

In the Academy of Sciences lecture, Tesla again found it appropriate to present in some detail the techniques which afforded him such success by describing a process of insulating condenser and inductor components in a manufacturing sense. This method was patented by him less then two months before the lecture and consists of excluding air or gas from the dielectric separating conductors of high potential in close proximity, or remote portions of the same conductors, in as perfect a manner as possible and in a convenient and practicable way.[23] In later years, all rolled-foil condensers having waxed-paper dielectric were manufactured by a modification of the method Tesla describes and pioneered by the Cornell Dubilier Company.

With figure 11, Tesla diagrammatically shows a pair of secondary coils wound with turns of each aiding—their mid-point connection brought out as a tap, the entire unit designed to withstand the encapsulation pressure exerted by the manufacturing process previously described. Tesla here makes a significant statement concerning the length of secondary coil windings, that each section be equal to or somewhat less than a quarter wavelength of electromagnetic propagation in the winding, thus allowing a maximum potential difference at the terminals. As such, these coils represent a half-wave pair.

Tesla goes on to describe the functioning of the spark gap in a primary circuit, critically examining the early investigations of Johann Poggendorff (1786-1877) on the action of a

[23] U.S. Patent No. 577,671 of Feb. 23, 1897, "Manufacture of Electrical Condensers, Coils, &c.," application filed Nov. 5, 1896.

break in a vacuum,[24] and suggests various improvements in the design of the break in the primary circuit. He describes the necessary attention to the construction of the break to produce currents of high voltage and frequency "causing showers or continuous streams of thick, thundering sparks to dart out into space to a distance of 8 or 9 feet ... sometimes like veritable lightning bolts." Many readers today, in looking at the photographs of the discharges that Tesla produced in his laboratories, usually do not consider the din associated with such displays. Tesla has made reference in other writings to wearing ear plugs, and it has been reported that the noise from his experimental station at Colorado Springs could be heard 10 miles away.

An element of "the discoverer's keen searching sense" is found in his observing an "exaggerated Thomson effect."[25] He noticed that the primary circuit of heavy copper ribbon, the inactive portion of which should be no more than five percent, is perceptibly cooler in extended working, heat apparently carried away from the primary ribbon to the foil coatings of the condenser. Of the thousands of Tesla coils constructed since then, no reports have appeared in the open literature on this subtle, difficult-to-measure effect.

Another aspect of the primary circuit, as was found for the secondary circuit, is critical attention to its length. Tesla describes that as little as a quarter of an inch change in length of the primary circuit will have a pronounced effect on the performance of a coil! Tesla advocates that the best result is attained when a stationary wave is formed with a single

[24] Poggendorff, J., "Effects of Interrupting a Current Within Discharge Tube," *Philosophical Magazine*, 4th sec., Vol. 10, 1855, pp. 203-307.

[25] *Encyclopaedia Britannica Micropaedia* (1982) defines the Thomson effect as "the evolution or absorption of heat when electric current passes through a circuit composed of a single material that has a temperature difference between two points along its length. This transfer of heat is superimposed on the common production of heat by currents flowing through conductors because of their electrical resistance. If a copper wire carrying a steady electric current [i.e., direct current] is subjected to external heating at a short section while the rest remains cooler, heat is absorbed from the copper as the conventional current approaches the hot point, and heat is transferred to the copper just beyond the hot point."

node located at a point of the discharge circuit or conductor equidistant from the opposite condenser coatings, as illustrated below.

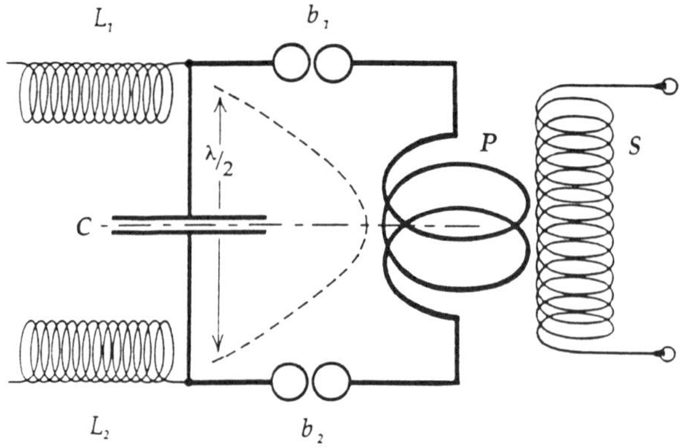

With this design, a half-wave primary circuit length results. This may be difficult to achieve in practice for the construction of Tesla coil systems as revealed by the table shown below. For the desk-size unit illustrated in figures 9 and 12, for example, the operating frequency would be in excess of 50 Mc.

Operating frequency of secondary	Length of primary to achieve half-wavelength
10 kc	9.3 mi
50 "	1.9 "
100 "	4,900 ft
500 "	984 "
1 Mc	490 "
5 "	98 "
10 "	49 "
50 "	10 "

It is obvious that for a secondary coil constructed as above, and the primary constructed according to the same design criterion as described earlier, both coils will have nearly the same length and, essentially, the same number of turns—a 1:1 turns ratio. It might at first be asked, "How do we obtain any voltage gain?" Voltage gain in Tesla coil systems is not achieved solely from the secondary/primary turns ratio; the Q of the coils and a standing-wave condition that results in coils of appropriate length also contribute substantially to the gain. Although significant resonant rise can be achieved in a transformer having a 1:1 turns ratio from high Q values and properly adjusted length of primary and secondary coils, such design is rarely undertaken.

The desk-size units illustrated in figures 9 and 12 require careful design of the primary circuit, operating from municipal supply circuits of 110 volts, to obtain high current excitation of the primary coil. A half-wave circuit loop from the condenser coatings is a unique way to accomplish this. But, when a high-voltage, high-current supply is available, then the advantage of higher numerical values for secondary/primary turns ratio (approximately the square root of the respective inductance values ratio) prevails.

Tesla advocated that the length of a grounded secondary coil should be a quarter wavelength of the oscillation frequency, yielding the maximum potential at its terminal by virtue of a standing wave condition. The first hint of this design technique was a photograph appearing in the May 20, 1896 issue of the New York *Electrical Review* showing Tesla seated beside an 8-foot diameter flat-spiral coil in his laboratory on Houston Street but without accompanying explanation. During this lecture, Tesla provided two diagrams showing these flat-spiral coils in experimental wireless transmitting and receiving antenna circuits, again without explanation.[26] But following this lecture, Tesla applied for his first patent on wireless telegraphy showing such coils and describing the technique, now universally employed, for quarter-wavelength radio transmitter antenna circuits.[27]

Oliver Shallenberger, known for his development of the induction ampere-hour meter in 1888 for the Westinghouse Electric and Manufacturing Company, which had committed itself to alternating current power equipment production by

[26] See Sec. I-Addendum, Figs. 15 and 16, pp. 72, 73.

[27] See note 4.

acquisition of the "famous 30" Tesla patents on motors and systems, contributed information that has not appeared elsewhere concerning the high frequency, high voltage apparatus used by Tesla in obtaining the results presented in the first of his 1896 X-ray communications to the *Electrical Review.*[28]

> "As to the apparatus and operation, which are especially important, it may be stated that the current was taken from an alternator, of a frequency of 133 periods per second, and passed through a primary coil of a transformer for increasing the e.m.f. from 100 volts to from 16 to 25 thousand. The secondary current was then passed through Leyden jars and a double cascade of slightly separated brass cylinders, whereby it was changed into an oscillatory current of an extremely high frequency, which was then connected through the primary of a second induction coil having very few turns of wire, and no iron core and having a ratio of 7 to 1. By this means the e.m.f. was raised to somewhere between 160,000 volts to 250,000, and was used to energize the discharge tube for the generation of X rays."

Thus we find that for driving experimental evacuated tubes, Tesla employed resonant-coil apparatus preserving a fractional wavelength for the secondary, rather than the primary.

The determination of exact operating frequency of oscillators was a vexing problem for all early investigators, a difficulty that those today working with oscilloscopes having linear time-base sweeps may not fully appreciate. Wave meters, essentially a closed oscillatory circuit, usually with calibrated substitution inductances and variable condensers, were used in the near years of WWI.[29] But before the turn of the century, experimenters and researchers devised their own ad-hoc methods of frequency determination.

In this lecture, Tesla describes a carefully calibrated device that he did not push to patent but for which the diagrams

[28] Thompson, E.P., *Roentgen Rays—and the Phenomena of the Anode and Cathode* (D. Van Nostrand, 1896), pp. 136-138.

[29] See, for example, Mauborgne, J.O., *Practical Uses of the Wave Meter in Wireless Telegraphy* (Mc-Graw Hill, 1913).

presented in the lecture clearly indicate that a patent was considered. The patentable aspect of the device was the "visual synchronism" capability in reading divisions or markings of a uniformly rotating disk illuminated with periodic flashes from neon or spark sources associated with the system to be measured. This device is recognized today as the electronic-flash stroboscope pioneered by Harold E. Edgerton (1903-1990).

The mechanical stroboscope fork, a tuning fork with slit shutter, had been used by others unaware of Tesla's invention until about 1910 for uniform motion measurements. When this editor showed Dr. Edgerton a copy of a portion of this lecture revealing Tesla's much earlier work with stroboscopes using neon and arc flash illumination, Edgerton replied making no claim of invention of the concept.[30]

The disk shown in Tesla's drawing Fig. 14 has 10 radial segments on the outer ring, with radial segments of 9, 8, 7, and 6 for the succeeding inner rings. Note that in Fig. 14, the outer ring is further subdivided into 10/20 divisions per segment with a vernier adjacent to that ring.

Let **r** be the number of revolutions per second, **s** the number of radial segments in the ring, and let **a** and **b** have integral values (**1**, **2**, **3**, etc.). If the frequency of the flashes differs from **r**(**s**·**a**/**b**) by **1**/**b** cycles per second, then **a**·**s** radial segments will appear to revolve at a rate of 1 radius segment per second. Thus, a disk of **s** radial segments illuminated with **b**·**s** flashes during each revolution appears like a disk having **b**·**s** segments, and if the flashes occur **s**/**b** times for each revolution, the disk appears to have **s** radial segments.

For this method to be used as a detection scheme, it is clear that the presence of continuous-wave signals would not be of sufficient strength to excite a small neon flash lamp, typically requiring 90 volts for ignition. Tesla later describes biasing the circuit containing the lamp to near ignition with a toy battery pack. A number of these detection circuits are to be found in Tesla's telemechanics patent and in a pair of

[30] Personal communication, Feb. 22, 1988.

means and methods detection patents issued during the succeeding four years.[31]

The method of synchronous rotating disks or spindles was employed by Tesla in continuous-wave detection devices and presented in three patents issued from 1901 to 1905.[32] It was the discussion of such receiving methods in the lecture together with visual synchronization measurement and detection schemes, the publication of which would preempt patent disclosures, that undoubtedly caused Tesla not to commit this portion of the lecture to publication.

Lenard and Roentgen rays

The discovery of X rays was directly associated with advances in achieving high degrees of vacuum, following the discovery of the beautiful visual effects produced by William Crookes with partially evacuated tubes containing various residual gas compositions and phosphor coatings. At pressures of a few centimeters of mercury, a shimmering pink streamer is observed along the length of the tube which fills the entire cross section as the pressure is further reduced. At about 2.5 millimeters, this glow begins to separate into identifiable regions until reaching 0.1 millimeters, and proceeding from the cathode at one end of the tube to an anode at the other end, the familiar patterns now known as

[31] U.S. Patent Nos. 613,809 of Nov. 8, 1898, "Method and Apparatus for Controlling Mechanism of Moving Vessels of Vehicles," application filed July 1, 1898; 685,954 of Nov. 5, 1901; "Method of Utilizing Effects Transmitted Through Natural Media," application filed Aug. 1, 1899; and 695,956, "Apparatus for Utilizing Effects Transmitted Through Natural Media," application filed Nov. 2, 1899.

[32] U.S. Patent Nos. 685,953 of Nov. 5, 1901, "Method of Intensifying and Utilizing Effects Transmitted Through Natural Media," application filed June 24, 1899; 685,955 of Nov. 5, 1901, "Apparatus for Utilizing Effects Transmitted From a Distance to a Receiving Device Through Natural Media," application filed Sept. 9, 1899; and 787,412 of April 18, 1905, "Art of Transmitting Electrical Energy Through the Natural Mediums," application filed May 16, 1900.

the Crookes dark space (in German literature, Hittorff dark space), negative glow region, Faraday dark space, and lastly the positive column with striations are observed. It is the positive column that is seen in neon signs operating at several millimeters of pressure.

Lowering the pressure from 0.1 millimeters of mercury, the Crookes dark space lengthens with the striations in the positive column becoming farther apart. At roughly 0.08 millimeters, the cathode dark space reduces in length to about 1 centimeter. The Crookes dark space, negative glow, and Faraday dark space regions will remain fixed in length along the tube with the positive column taking up the remaining length of the tube regardless of the tube's length. At pressures substantially below 0.01 millimeter, a green fluorescence appears on the inner walls of the tube, and below 0.001 millimeter, the tube becomes dark.

As an aid in evaluating the degree of vacuum obtained for the Lenard and Roentgen tubes that Tesla investigated and demonstrated, when a vacuum is on the order of 0.001 millimeter of mercury the X rays are easily absorbed, minimally passing through the human hand. These rays are referred to as *soft* X rays, roughly a few angstroms in wavelength. At 0.0003 millimeters of mercury, exceedingly high voltages are needed to produce a discharge and the X rays are quite penetrating, able to pass through the bones of a hand with little absorption. These rays are referred to as *hard* X rays, about 0.1 angstrom in wavelength.

It is significant that Tesla considered both Lenard and Roentgen rays in his communications and lecture regarding "penetrating rays" through substances. Some astonishment was expressed by radiologists of the results Tesla achieved in his experiments not only at the time they were presented in the lecture and in his 1896-97 communications appearing in the *Electrical Review* but as well by those reviewing these reports many years later. This is primarily because Tesla's researches involved *both* Lenard and Roentgen apparatus, as the titles of his reports so state, but his communications on the subject have generally been interpreted in terms of only X-ray apparatus and effects.

In briefly reviewing the "rays" topic, it is noted that Philipp Lenard (1862-1947), in 1893, announced the discovery of invisible rays produced in a Crookes tube and capable of passing through a thin aluminum window. Now known as "Lenard rays", these are cathode electron beams capable of passing several centimeters beyond the window in air. The thin aluminum window has also become known as a "Lenard window," and in his experiments penetrating rays were produced at the point of first impact—the window.

Neither Lenard, Roentgen, Tesla, or other researchers knew what they were dealing with in terms of "rays" up to that time. They, as well as other investigators, were at the forefront of investigation, undertaking researches to answer the fundamental questions of the nature of rays and particles. In the Academy of Sciences lecture, Tesla demonstrated a source of powerful rays which he described as more powerful than any before available.[33] The source of the rays was an arc between closely-spaced electrodes in vacuum, now recognized as extreme ultraviolet radiation (approx. 500 angstroms). The ability to distinguish *soft* X Rays and extreme ultraviolet rays was then difficult.

But 1897 marked a new era in physics with the identification of the sub-atomic particles. That year, Joseph Thomson identified electrons by measuring the ratio of their charge to mass, e/m, and knowing the charge the mass could be calculated.

Lord Kelvin was in opposition to the evolving theories of atomic structure advanced by Thomson, Rutherford, and others. He was greatly influenced, for most of his lifetime, by the ideas of Rudjer Boscović (1711-1787), an extraordinarily remarkable scientist who engaged in engineering, architecture, and archeology.[34] Of Boscović Tesla remarked in an unpublished 1936 interview article,

[33] Refer to Appendix for a review of the lecture appearing in the *Electrical Engineer*, April 14, 1897.

[34] A week-long celebration examining the life and work of Boscović was held in Rome, May 1988, in observance of the 200th anniversary of his death.

> "The relativity theory by the way, is much older than its present proponents. It was advanced over 200 years ago by my illustrious countryman Boscović, the great philosopher who, not withstanding other and multifold obligations, wrote a thousand volumes of excellent literature on a vast variety of subjects. Boscović dealt with relativity, including the so-called time-space continuum...."

Kelvin, in his 1884 Baltimore lectures,[35] referred to Boscović's great book,[36] and in the 1890s adopted Boscović's interpretation of forces, the "force curve." When Joseph Thompson published the discovery of the electron particle in 1897, Kelvin extended the concepts of Boscović to systems of electrons accounting for all electrical phenomena and radioactivity—a model explaining electrons ejecting at high velocity. Although not referenced in Tesla's contemporary writings on X rays, but acknowledged years later, the theories of Boscović and Kelvin had great influence on his interpretation of the experimental results he obtained.

Let us examine some of the specific results described by Tesla in his researches during the 1896-97 period. Tesla correctly concludes that the source of penetrating rays is "the place of *first* impact of the projected stream of particles within the bulb."[37] In the familiar X-ray tubes of today, the rays emanate from a massive anode target inside a tube bombarded by an emission stream of electrons from a heated cathode under very high potential difference. The anode target is, in this case, the place of first impact of particle streams produced inside the tube.

In his communication "On Reflected Roentgen Rays," appearing in the April 1, 1896 *Electrical Review*, Tesla gives a

[35] *Baltimore Lectures on Molecular Dynamics and the Wave Theory of Light* (London: Cambridge University Press, 1904).

[36] *Theoria Philosophiæ Naturalis redacta ad unicam legem virium in natura existentium* (Venetia: Editio Venetia prima ipso auctore præsente, et corrigente, 1763).

[37] See Section III.

qualitative description of the intensity of rays which he interpreted to be reflected from a dozen metal and nonmetallic surfaces. He also reported that the relative strength of the re-radiations from the various metals agreed with the sequence of these metals in the activity series developed by Volta. At that time, the phenomenon of secondary radiation of X rays had not been hypothesized. It was not until 1922 that Arthur Compton presented in monograph form a definitive analysis of secondary X-ray radiation.[38] It showed that Tesla's series of the relative strength of what he interpreted to be "reflected rays" from various metals agreed with the mass absorption coefficients for these metals when excited by soft X rays; i.e. in the region of 1 angstrom. The following year, the first successful reflection of X rays with a very small incident glancing-angle of the radiation was reported by Compton, thus revealing the experimental difficulty.

Tesla found no evidence of X-ray diffraction, but a dozen years later research apparatus became available providing W.H. Bragg, Max von Laue, Ernst Pohl, and Bernhard Walter the opportunity to show diffraction. As well, Tesla found no evidence of refraction. In 1925 the first successful experiments showing refraction were reported by Larsson, Siegrahn, and Waller.[39]

A review of the Academy of Sciences lecture finds Tesla stating he had succeeded in deflecting Roentgen rays by a magnet—the rays also charging a condenser at some distance.[40] But in his communications appearing in the April 22 and August 12, 1896 issues of the *Electrical Review*, he modifies his remarks saying Lenard rays could be deflected by a magnetic field whereas Roentgen rays could not as observed by exposure of fluorescent-emulsion films. Thus, he identified a distinction between them in their producing penetrating rays. When the energy of Lenard rays is abruptly

[38] Compton, A.H., "Secondary Radiations Produced by X-Rays," *Bulletin of the National Research Council,* October 1922, (Vol. 4, Pt. 2, No. 20), the third and last of a series which formed the report of the Committee on X-Ray Spectra.

[39] Larsson, A., Siegrahn, M., and Waller, T., "The refraction of x rays," *Physical Review*, Feb. 1925, p. 235.

[40] Refer to Appendix for a review of the lecture appearing in the *Electrical Engineer*, Apr. 14, 1897.

decelerated by a magnetic field, either from a bulk external source or an individual atom, as in passing through a Lenard window or in the bombardment of a massive target, the phenomenon of bremsstrahlung radiation results.

Bremsstrahlung is observed only for light particles such as electrons moving with high energy through matter. The particles "spit radiation in a forward direction;" that is, the direction of motion.[41] From Tesla's descriptions of work with Lenard apparatus, X rays produced by bremsstrahlung is an explainable result.

The tubes that Tesla employed in his researches had only one electrode, a cathode. He earlier remarked,

> "Clearly, if we put two electrodes in a bulb ..., we limit the potential, for the presence not only of the anode but of any conducting object has the effect of reducing the practicable potential on the cathode."[42]

Thus, for such a tube, the place of first impact of an emission stream emanating from the cathode would be the glass envelope at the other end of the tube.

In his communication appearing in the March 18, 1896 *Electrical Review*, Tesla describes obtaining X-ray shadow graphs at a remarkable distance of 40 feet. Even if it assumed that X rays were produced at the glass envelope at the end of the tube, that would not fully account for the ability to obtain X-ray shadow graphs at such a great distance as 40 feet.

Tesla's researches were, in the main, with Lenard rays —electron rays passing through a thin aluminum window, and with beams of particles that could not be distinguished at that time as electrons or molecular particles. He gives a most interesting description of the "warming-up" process in working a tube to the point where penetrating rays are copiously produced:

> "If we attach a fairly exhausted bulb containing an electrode to the terminal of a disruptive coil, we observe

[41] Feynman, R.P., *Lectures on Physics* (Reading, Mass.: 1963, Addison-Wesley), Vol. 1 of 3, p. 34–6.

[42] *Electrical Review*, March 11, 1896.

> small streamers breaking through the sides of the glass. Usually such a streamer will break through the seal and crack the bulb, whereupon the vacuum is impaired; but, if the seal is placed above the terminal, or if some other provision is made to prevent the streamer from passing through the glass at that point, it often occurs that the stream breaks out through the side of the bulb, producing a fine hole. Now, the extraordinary thing is that, in spite of the connection to the outer atmosphere, the air cannot rush into the bulb as long as the hole is very small. The glass at the place where the rupture has occurred may grow very hot—to such a degree as to soften; but it will not collapse, but rather bulge out, showing that a pressure from the inside greater than that of the atmosphere exists. On frequent occasions I have observed that the glass bulges out and the hole, through which the streamer rushes out, becomes so large as to be perfectly discernible to the eye. As the matter is expelled from the bulb the rarefaction increases and the streamer becomes less and less intense, whereupon the glass closes again, hermetically sealing the opening. The process of rarefaction, nevertheless, continues, streamers being still visible on the heated place until the highest degree of exhaustion is reached, whereupon they may disappear. Here, then, we have a positive evidence that matter is being expelled through the walls of the glass."[43]

This curious process, in its examination a near century later, suggests that the internal "pressure" was actually the result of an internal force. Tesla operated his single-electrode bulbs at extremely high voltage, high frequency currents. An electron beam develops at the cathode as a result of high-field emission during the negative half of the alternating current cycle.[44] It concentrates on a point defect at the other end of the tube arising from either localized ionic stress, trace impurities, or a high concentration of dopant additives. The

[43] *Electrical Review*, March 18, 1896.

[44] For a discussion of the internal process of single-electrode bulbs excited by high frequency, high voltage alternating currents, see Corum, J.F., and Corum, K.L., "Critical Speculations Concerning Tesla's Invention and Applications of Single Electrode X-Ray Directed Discharges for Power Processing, Terrestrial Resonances and Particle Beam Weapons," *Proceedings of the 1986 International Tesla Symposium*, Colorado Springs, pp. 7-21 - 7-44.

spot then becomes a virtual anode because of its temperature increase, between annealing and melting points (variable with glass composition, approx. 450-500°C and 1,200-1,500°C, respectively), resulting in significant conductivity to the outer surface in contact with air—a dispersed-medium return loop of the high voltage source impressed on the cathode. When a point reaches redness in color, its specific resistance can drop from 100 trillion ohm-centimeters at room temperature to 10,000 ohm-centimeters at the annealing temperature and unity at the melting point. In using such extremely high voltages on Lenard and single-electrode tubes, Tesla advises in his communication appearing in the March 18, 1896 issue of the *Electrical Review* not to overheat them in continued use.

This editor has viewed a video of an experiment performed in 1985 by Eric Dollard showing the same process of a single-electrode bulb developing a hot spot, the glass bulging out, rarefaction increasing, and the bulb subsequently resealing. This process has yet to be critically investigated.

Tesla's December 2, 1896 communication to the *Electrical Review* refers to a "material stream" projecting from the single-electrode bulb and he later remarks,

> "Roentgen gave us a gun to fire—a wonderful gun, indeed, projecting missiles of a thousandfold greater penetrative power than that of a cannon ball, and carrying them probably to distances of many miles, with velocities not producible in any other way we know of."

This germinal idea of projected particle beams in air, and the succeeding experimental work undertaken in Colorado Springs in 1899 utilizing extremely high potentials and the production of relativistic electron beams, undoubtedly crystallized Tesla's design approach for a particle-beam weapon which he presented to the allied powers as WWII stormclouds were gathering.[45]

[45] This design approach appears in *Proceedings of the Tesla Centennial Symposium*, Colorado Springs, 1984, pp. 144-150, using an abbreviated form for unit expressions; e.g., acceleration expressed simply as "meters" rather than "meters/second^2." A copy of this design approach is as well held by the Tesla Museum in Belgrade.

Harmful actions from Lenard and Roentgen tubes

The portions of the lecture that Tesla did choose to publish were in the form of two communications appearing in the May 1 and August 9, 1897 issues of the *Electrical Review* (Secs. II and III of this reconstructed lecture) on the subject of deleterious actions from Lenard and Roentgen tubes . He felt an urgent necessity to present warnings to the medical community and those undertaking research in penetrating rays of the possible dangers in the operation of apparatus for their production.

Tesla describes radiation experiments that today would be unthinkable to subject the body—exposing a hand to radiation from a Lenard tube and the hand becoming warm and later much reddened and swollen. Believing this may have been the result of a mechanical injury, he again thrust his hand close to the window of the tube and instantly felt pain. The pain lasted a few days afterward and later he observed that all the hair was destroyed and that the nails on the hand had grown anew. He describes the tightening of skin, or stiffening of muscles, when a hand is held close to the window of the tube. In a sever case, the skin gets deeply colored and blackened in places, and ugly, ill-foreboding blisters form; thick layers come off, exposing raw flesh which discharges freely. Burning pain, feverishness followed.

When conducting research into unknown realms involving the human body, pioneers, who either inadvertently or with known potential risks undertook experiments having uncertain outcome to themselves, taught the valuable lessons from which we benefit today. As a result of the harmful effects Tesla experienced, he continues in these communications the advocacy for the proper construction and shielding of the apparatus with particular concern for medical professionals and patients.

Tesla was the constant nemesis of Elihu Thomson, an electrical inventor and manufacturer of the period. Neither Thomson nor the Thomson-Houston Electric Company (later acquired by the General Electric Company) were successful in challenging Tesla or the Westinghouse Electric and

Manufacturing Company on the Tesla patents for alternating current power distribution systems and motors. The scene in courts was all too familiar—Tesla walking in, dressed in formal attire accompanied by his attorneys, and as a witness astonishing the court and observers with a remarkable display of recall and caustic wit.

A tawdry impress on an overzealous biographer shows in attempting to elevate the subject by misstatements of fact concerning the subject's competitor or adversary. Thomson's biographer, David Woodbury,[46] saw in the Columbian Exposition demonstrating Tesla's system of alternating current power distribution a nonevent—Thomson presumably had also accomplished the various exhibited demonstrations!

In an instance involving Thomson and X rays, Harold Abrahams and Marion Savin, in their compilation of Thomson's correspondence,[47] engage in unbecoming historiographic distortion by entering a surprisingly impertinent reference note for a letter from Dr. William Greene to Thomson dated December 20, 1896, mentioning an X-ray burn suffered by Thomson on his finger during experiments. The editors mention Thomson's "lively controversy with Nikola Tesla who thought that X rays were harmless...." No record exists of such an exchange on the subject of harmful effects of X-ray radiation, but a lively exchange did occur between them six years prior to the Academy of Sciences lecture on the nature and effects of high frequency currents. Although beneficial for the technical readership in providing an airing of the issues, it appeared at the expense of Thomson.[48]

The reconstructed Academy of Sciences lecture follows. It is now seen as a contribution to the history of the emerging scientific discoveries and technological developments of that period not previously presented.

L.I.A.

[46] Woodbury, D.O., *Beloved Scientist* (McGraw-Hill, 1944).

[47] *Selections from the Scientific Correspondence of Elihu Thomson* (see note 8).

[48] The exchange occurred in a succession of communications to the *Electrical World* following the publication in its Feb. 21, 1891, issue of the first of Tesla's *trio-series* lectures on high frequency alternating currents: Thomson, Mar. 14, pp. 204-5; Tesla, Mar. 21, pp. 223-4; Thomson, Apr. 4, p. 254; Tesla, Apr.11, pp. 272-3.

I haunted thee where the Ibis nods,
From the Bracken's crag to the Upas Tree.

N. Tesla
November 4, 1934

Section I

IMPROVED APPARATUS FOR THE PRODUCTION OF POWERFUL ELECTRICAL VIBRATIONS; NOVEL HIGH FREQUENCY MEASUREMENT METHODS.

Ladies & Gentlemen:

You will still remember vividly, no doubt, the excitement which a year ago was caused by the announcement of the discoveries of Professor Roentgen. Suddenly, without any preparation, Roentgen surprised the world with two wonderful results. He showed us how to take a photographic impression of an object invisible to the eye and, what seemed more extraordinary, he enabled us, by the help of his luminous screen—now known as the fluoroscope—to see, with our own eyes, the outlines of the object. We are living in an age of exceptional intellectual activity, and important advances are often recorded, but these were almost of the order of the telescope and microscope and such discoveries come no more than once or twice in a century. Scarcely can any one of us hope to again witness in his lifetime an event of so widespread a scientific and popular interest. The desire to see things which seem forever hidden from sight is more or less strongly developed in every human being,through all degrees of this sentiment, from the idle curiosity of the unenlightened to the absorbing desire for knowledge of the highly refined, and this in itself was sufficient to engage universal attention; but, apart from this, these discoveries brought promise of relief to numberless sufferers and stirred all over the world the fibers of humanity. It is hardly necessary for me to tell you that the fever took hold of me also, but mine was a singular, grave case, and I had not recovered from its effects to this day. I hope you will pardon here a slight digression which I have a strong reason to make.

and not so much those of incandescent vacuum tubes, although some photographs were likewise taken with these. As both the artists and myself were busy on other matters the plates in their ordinary holders were frequently put in some corner of the laboratory until a suitable opportunity for carrying on the experiments was found. During these investigations many plates gave a result, while many others failed, and on some of these both Mr. Alley, who then assisted me, and myself noted unaccountable marks and defects. Mr. Alley particularly found it extraordinary that, in spite of his care, many plates proved defective and unsuccessful. The taking of these photographic impressions by means of Crookes bulbs brought freshly to my mind the experiments of Lenard, some features of which, particularly the action on a sensitive plate, had fascinated me from the start, and I resolved to go over the ground covered by him with assistance and improved appliances. Just as my attention was arrested by this feature my laboratory with almost everything it contained was destroyed; and the few months following passed in intense activity which made me temporarily forget my projects. I had hardly finished the work of reconstruction and resumed the course of my ideas when the news of Roentgen's achievement reached me. Instantly the truth flashed upon my mind. I hurried to repeat his incompletely reported experiments, and there I beheld the wonder myself. Then -too late- I realized that my guiding spirit had again prompted me and that I had failed to comprehend his mysterious signs.

(3)

Reproduction of lecture text page.

At the close of 1894, realizing the necessity of recovery from a straining task, on which I have been laboring for a number of years and which still commands my energies, it occurred to me to investigate the actinic action of phosphorescent bodies. The subject did not appear to have been studied, and I began the work at once, securing later, at the suggestion of some friends connected with the *Century Magazine,* the assistance of Messrs. Tonnelé & Company, artists' photographers, of this city, then doing work for this magazine. In these experiments, I employed an improved apparatus for the production of powerful electrical vibrations as well as one of my high frequency alternators of old design. A great variety of Crookes tubes, single-electrode globes, and vacuum bulbs without external electrodes were experimented upon. A surprising fact was soon brought to light; namely, that the actinic power of the Crookes bulbs varied greatly and that some, which emitted a comparatively strong luminosity, hardly showed an effect, while others, of much smaller light-giving power, produced strong impressions. I wish to state here, in order to be clear, that my efforts were directed toward investigating such actions of true phosphorescent light, as furnished from bulbs without appreciable emission of heat, and not so much those of incandescent vacuum tubes, although some photographs were likewise taken with these. As both the artists and myself were busy on other matters, the plates in their ordinary holders were frequently put in some corner of the laboratory until a suitable opportunity for carrying on the experiments was found. During these investigations many plates gave a result, while many others failed, and on some of these both Mr. Alley, who then assisted me, and myself noted unaccountable marks and defects. Mr. Alley particularly found it extraordinary, that, in spite of his care, many plates proved defective and unsuccessful. The taking of these photographic impressions by means of Crookes bulbs brought freshly to my mind the experiments of Lenard, some features of which, particularly the action on a sensitive plate, had fascinated me from the start, and I resolved to go over the ground covered by him with assistance and improved appliances. Just as my attention was arrested by this feature, my laboratory with almost everything it contained was destroyed; and the few months following passed in intense activity which made me temporarily forget my projects. I had hardly finished the work of reconstruction and resumed the course of my ideas when the news of Roentgen's achievement reached me. Instantly the truth flashed upon my mind. I hurried to repeat his incompletely reported experiments, and there I beheld the wonder myself. Then—too late—I realized that my guiding spirit had again prompted me and that I had failed to comprehend his mysterious signs.

The statement of these facts might have been misinterpreted at the time of Professor Roentgen's announcement, and I have kept silent, although I was unable to overcome entirely my feeling in the introductory lines of my first of a series of articles I wrote on this subject in the columns of the *Electrical Review*. Presently, however, I have no fear of a misunderstanding of my works, and I am recording my painful but stimulating experience solely to make some of those, who have lightly written about the history of this new art, more justly appreciate this new departure. I was quite well acquainted with the results of Lenard and naturally often thought of his beautiful and promising experiments, and yet the possibility of the plates being marked and spoiled by some action of the bulbs never presented itself to my mind. While some might see in this only an argument for my own shortsightedness, others, more kindly disposed towards me, will with myself, consider it rather a demonstration of Goethe's great words, which I will not repeat in the text, but which say that, what Nature does not want to reveal to one's mind, one cannot force it from her with screws and levers.

But while I have failed to see what others in my place might have perceived, it was always since my conviction, which is now firmer than ever, that I have not been forsaken by the kind spirit who then communed with me, but that, on the contrary, he has further guided me and guided me right in the comprehension of the nature of these marvelous manifestations. Perhaps, in bringing to your attention some new facts which I have since discovered in addition to those already announced, I may induce, at least some of you, to interpret these phenomena as I do. For fear, though, that I might miss my chief object this evening, I must ask your kind indulgence to dwell in a few works on the novel appliances which are exhibited here for your inspection. When I trace their origin, I find it clearly in my early recognition of the fact that an economical method of producing electrical vibrations of very high frequency was the key for the solution of a number of most important problems in science and industry. Insignificant as these machines may seem to you, they are nevertheless the result of labors extending through a number of years, and I can truthfully say that many times the difficulties which I have encountered in my endeavors to perfect them have appeared to me so great as to almost deprive me of the courage to continue the work. When the experimenter has to spend several years of patient effort only to recognize that a mere microscopical cavity or air bubble in the essential parts of this apparatus is fatal to the attainment of the result sought for by him; when he has to find that his machine does not perform well because a wire he uses is a quarter of an inch too long or too short; when he notes that now a part of his apparatus when in action will grow colder in an apparently inexplicable way, and next that the same part will get overheated, though to all appearance the conditions are unchanged; when he makes puzzling observations at every step and ordinary instruments and methods of measurement are not available, then his progress is necessarily slow and his energies are severely taxed. Finally, I am glad to say, I have triumphed over at least the chief obstacles, and nothing of any serious consequence stands now in the way of obtaining electrical oscillations of frequencies up to a few millions a second from ordinary supply circuits with simple and fairly economical appliances. What this means I need not discuss. It will be duly judged by those who have kept in touch with the development in this and allied fields. These machines you see are only a few of the types I have developed, and as they stand here they are chiefly intended to replace the ordinary induction coil in its numerous uses.

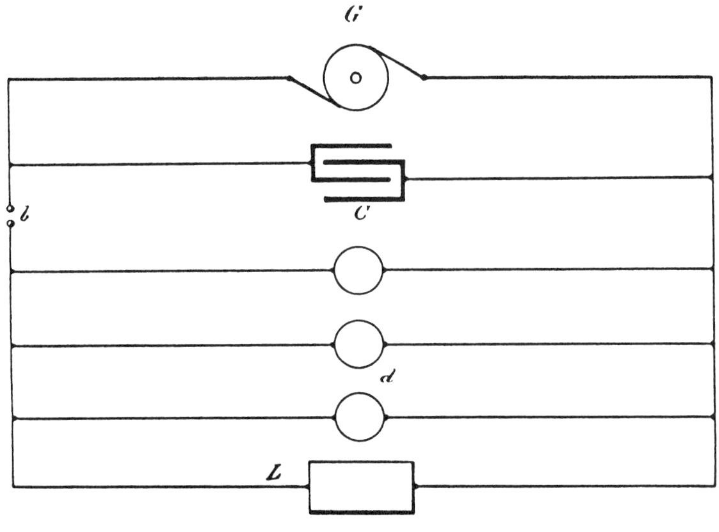

Fig. 1.—Method of transformation of electrical energy by oscillatory condenser discharges.

As to the broad principle, these transformers or electrical oscillators, as they might be most properly called, it is simple enough and has been advanced by me some five or six years ago. A condenser is charged from a suitable source and is then in any convenient way discharged through a circuit containing, as it does here, the primary of the transformer. The first diagram, Fig. 1, illustrates a generator *G*, a condenser *C*, and for charging and discharging the latter any kind of device *b* adapted to produce an intermittent break in the dielectric. The circuit *L* containing the high or low tension devices *d* through which the condenser discharges being properly adjusted, extremely rapid electrical vibrations which, so far we know are unattainable by any other means, result; and these set up, by inductive action in any neighboring circuits, similar vibrations which give rise to many curious phenomena. Having familiarized myself with these at the time when some laws governing them were not quite well understood, I have retained certain conceptions which I have then formed and which, though primitive, might stand even now in the light of our present advanced knowledge.

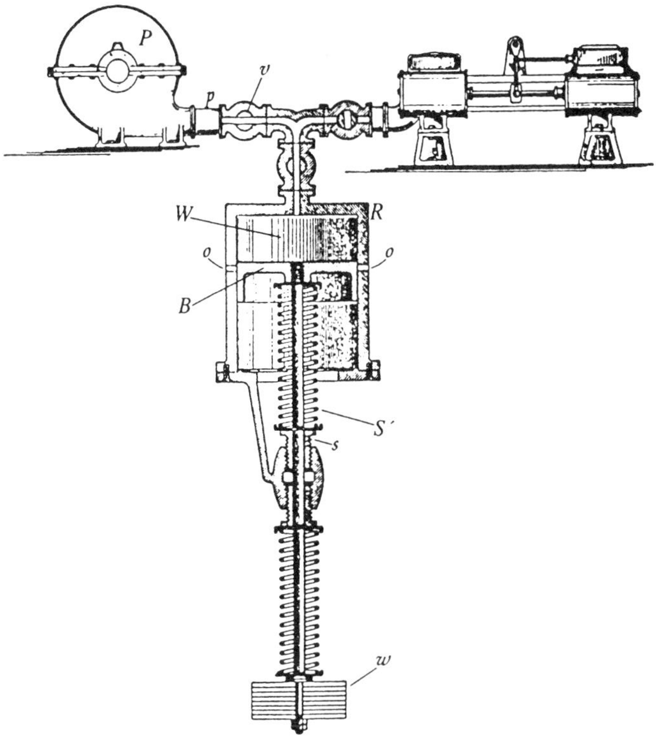

Fig. 2.—Mechanical analogy of electrical oscillator.

I have likened a condenser to a reservoir *R* into which by means of a pump *P* an incompressible fluid as water *W* is supplied through a feed pipe *p*, as illustrated in the second diagram, Fig 2, the fluid representing electricity, the pump the generator and the feed pipe the connecting wire. The reservoir has a movable bottom, held up by a spring *S*, and opens the ports *oo* when the fluid in the vessel has reached a certain height and the pressure has become sufficient to overcome the elastic force of the spring. To complete the model, adjustable weights *w*, a screw *s* for allowing the tension of the spring, and a valve *v* for regulating the flow of the fluid are provided. With the giving away of the bottom, the fluid in the reservoir acquires velocity and consequently momentum, which results in an increased pressure against the bottom causing the latter to open wider, and more of the fluid rushes out than the feed pipe can supply, whereupon the spring reasserts itself, closing again the ports, and the same process is repeated in more or less rapid succession. This opening and closing of the bottom may be likened to the making and breaking of the conducting path, the frictional resistance in this mechanical system to the ohmic resistance and, obviously, the inertia of the moving masses to the self-induction of the electric circuit. Now it is evident that, in order to keep in action the mechanism without the employment of auxiliary means, the average rate of supply through the pipe must be inferior to the average rate of discharge through the bottom; for, if it be otherwise, the ports will simply remain open and no vibration will take place. The more nearly the average rate of supply equals the average rate of discharge, the quicker will the bottom open and close; and it is furthermore clear from a consideration of simple mechanical principles that, if the fluid be supplied so fast through the feed pipe that the bottom vibrates as it would of its own accord, then the amplitude of the vibration will be the largest, the pressure against the bottom the strongest, and the greatest amount of fluid will be passed through the ports. All these considerations hold good for the electric circuit, and in experiments with high frequency machines, in which these effects were purposely magnified with the view of rendering their observation more easy, I have found that that condition is fulfilled when the capacity, self-induction, and frequency of vibration bear a certain relation, which observation I have since utilized in the adjustment of inductive circuits. You will note that this condition governing the rate of supply and discharge, most important in practice, especially when no positively acting mechanical means are employed for effecting the rupture of the dielectric, is a distinct one and should not be confounded with the condition determining the oscillatory character of the discharge investigated long ago by Lord Kelvin.

The next step in the evolution of the principle and its adaptation to practical uses was to associate with the system illustrated in Fig. 1 a self-induction coil *L*, as shown in diagram Fig. 3, which modified the action in many now well

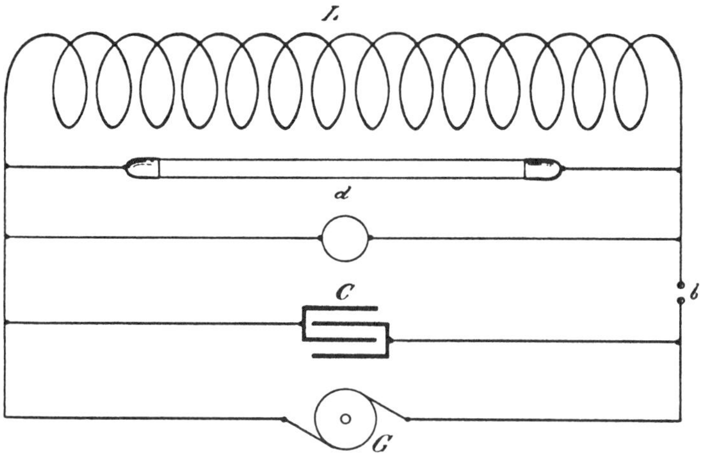

Fig. 3.—System illustrated in figure 1 with self-induction coil.

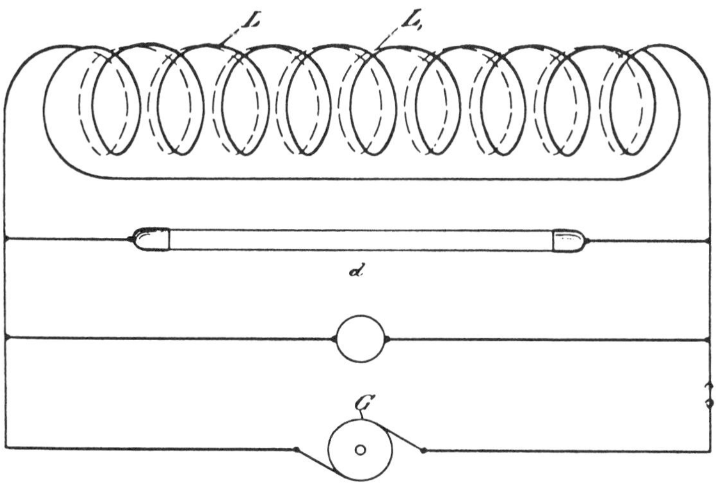

Fig. 4.—Coil wound to secure greatly increased capacity.

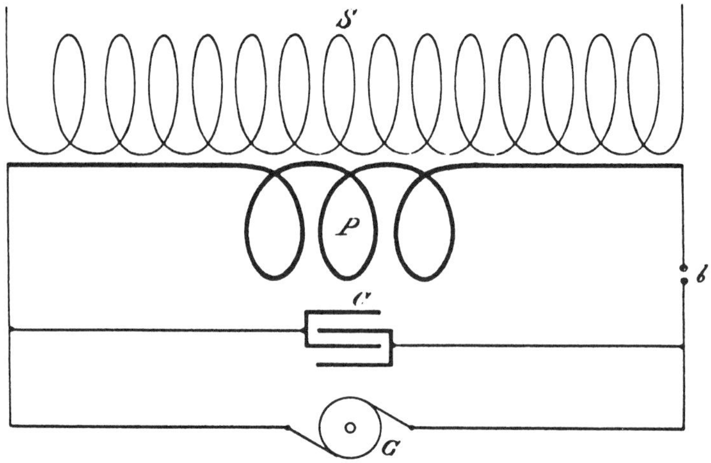

Fig. 5.—Associating a secondary coil with a primary circuit coil.

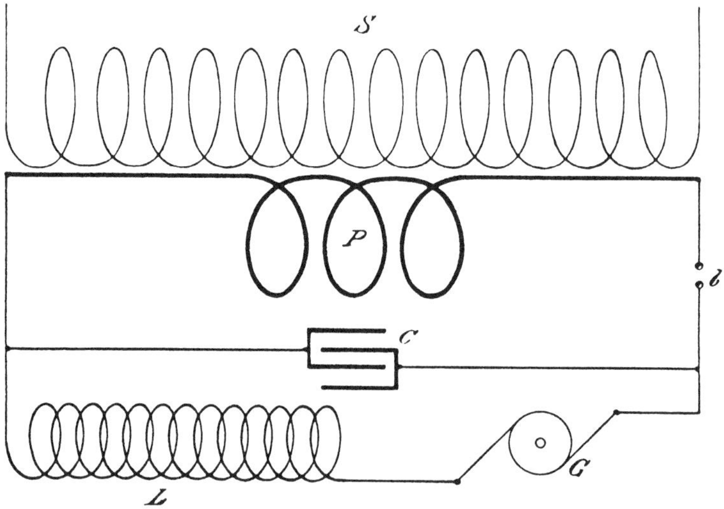

Fig. 6.—System adopted for existing municipal circuits.

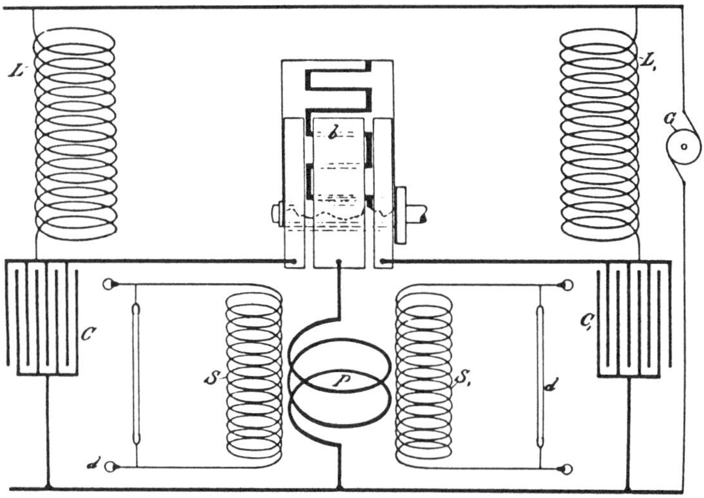

Fig. 7.—Circuit controller allowing condensers to discharge alternately and successively.

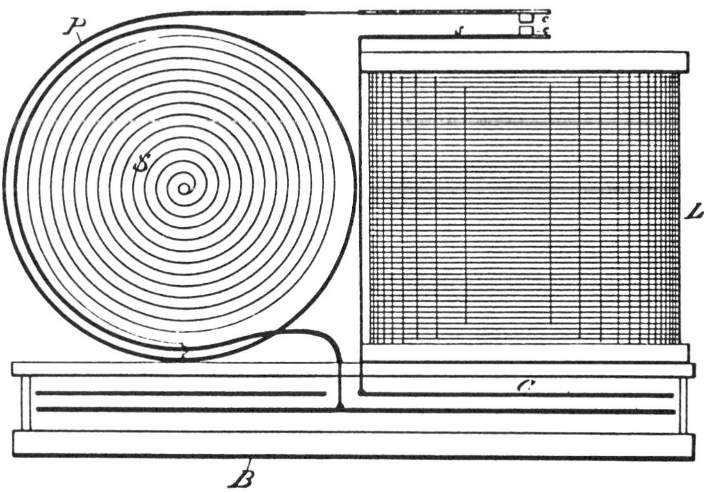

Fig. 8.—Arrangement of parts and circuits of a small oscillator.

understood ways. In a simplified form of this arrangement the condenser, as a distinctive part of the system, was done away with, the necessary capacity being given to the coil itself, and for this purpose the turns of the latter were wound as illustrated in Fig. 4 so as to allow the storage of the proper and generally the largest possible amount of energy. Then I associated a secondary coil *S* with the primary circuit *P*, as shown in Fig. 5, this enabling the obtaining of any tension required. After this, the arrangement in diagram Fig 6. was adopted as best suitable for the existing municipal circuits. Again, the self-explanatory diagram Fig 7. typically illustrates a further improved disposition as used in some of these machines with two or more circuits. A modification of this plan with one continuous contact common to the two circuits, and independent interrupters for each of these, allows easy adjustment of the phase of the currents through the primary, which is of practical advantage in some uses of the apparatus. Finally, in diagram Fig. 8 is shown the exact arrangement of the parts and circuits of one of these small oscillators with a break similar to that usually employed in connection with induction coils. Although the majority of the preceding arrangements have been described by me before, I thought it necessary to dwell on them here in order to present clearly and comprehensively the subject.

A specific result of value in the operation of Roentgen bulbs is obtainable by the use of two circuits linked as shown in Fig. 7, or otherwise, or entirely independent with two separate primaries. Namely, in the usual commercial bulbs the vacuum gets higher when the current is passed through the primary in a certain direction and is lowered when the direction of the current is reversed. This is a direct consequence of some conditions which, as a rule, are present in the operation of the usual apparatus; that is, the asymmetry of the opposite current impulses, the unequal size, configuration or temperature of the two electrodes, or like causes which tend to render unequal the dissipation of the energy from both the electrodes. It should be stated, though, that beyond a certain point, when the electrodes begin to act as entirely independent, the vacuum continues to increase no matter which way the current is passed through the primary. In the scheme illustrated in Fig. 7, or in its

Fig. 9.—Photograph of small oscillator diagrammatically shown in figure 8.

modifications referred to, the trouble attendant upon the operation of ordinary apparatus is practically done away with as the current though the primary is automatically reversed, and in this manner a tube which is first brought to the proper degree of exhaustion by means of one circuit can be worked for a long time without appreciable increase of vacuum or diminution of effectiveness.

A photograph of one of these finished instruments, Fig. 9, especially adapted to be used in the operation of Roentgen bulbs, or in general as a laboratory appliance in place of the ordinary induction coil, gives an idea of the actual arrangement of the parts. The condenser *C*, Fig. 8, is contained in a box *B* upon which is mounted in front the motor for controlling the circuits, in this instance simply a coil *L* actuating a spring *s*, fixed on top of the same. This coil, designated as the charging coil, serves at the same time to raise the pressure of the source to any value desired for charging the condenser. This is an important practical advantage, as it enables reduction of the capacity of the latter so that it need not be more than a few percent of that otherwise needed for an equivalent conversion of energy. Besides, the smaller the capacity, the quicker is the vibration and the shorter need be the high tension secondary. The discharge circuit *P* surrounding the secondary coil *S* is formed of a few turns of copper ribbon and mounted on the top of the box behind the charging coil, all connections being as short as possible so as to reduce as much as it is practicable both the self-induction and resistance of the discharge circuit. On the front side of the box, Fig. 9, containing the condenser, there are mounted the binding posts for connection with the line, two small fuses, and a reversing switch. In addition, two adjusting screws are provided for raising and lowering the iron core within the charging coil as a convenient means for varying within considerable limits the current of supply and regulating thereby the discharge of the secondary circuit. The instrument with rubber columns carrying the discharge rods, which are visible on the top, dismounted, can be enclosed in a box of 12 x 9 x 6 inches inside measure.

The mode of operation may be explained as follows: At the start, the spring contacts *cc*, Fig. 8, being closed and the condenser practically short circuited, a strong current passes through the charging coil attracting the armature fastened to the spring and separating the contacts. Upon this, the energy stored in the coil, assuming the form of a high tension discharge, rushes into the condenser charging the same to a high potential. The current through the coil now subsiding, the attraction exerted upon the armature ceases, and the spring reasserts itself and closes again the contacts. With the closing of the latter the condenser is discharged through the primary or discharge circuit, the constants of which are so chosen that an extremely rapid vibration of the electromagnetic system including the condenser and primary coil results. The currents of very high frequency thus obtained induce corresponding currents of high tension in the secondary. Simultaneously, however, with the discharging of the condenser, the current from the source of supply again rushes through the charging coil and energy is stored for the next charge of the condenser, this process being repeated as often as the spring opens and closes the contacts.

Although the instrument contains all the essentials of an ordinary induction coil, it will be seen that its action is entirely different, and the advantages of this new principle over the old are so great as to hardly require any lengthy comment. Merely to convey a true and more complete information I may mention a few of the most important ones. Take, for instance, the economy. The instrument referred to takes on a 110-volt direct-current circuit, according to load and adjustment, from 5 to 30 watts. It gives a powerful stream of sparks 6 inches in length, but if it be desired this distance can be easily doubled without increasing the energy consumed; in fact, I have found it practicable to produce by the use of this principle sparks of 1 foot in length involving no greater expenditure of energy than 10 watts. But in an in strument designed for a variety of uses, a departure must be made from a design insuring the greatest spark length. Of the total energy consumed by the apparatus, fully 80 percent can be obtained in the secondary circuit. Owing to the small total energy consumed and the efficiency of conversion, all parts of the instrument remain cool by long continued working with the exception of the contacts which, of course, are slightly heated. The latter are subject to much less deterioration than is commonly the case, as the condenser is small and, moreover, the current from the same does not, like in an ordinary coil, pass simply through the contacts and a few short connections, but has to traverse the primary coil, this reducing the current and diminishing very much the heating effects.

Consider next the advantages of the absence of fine wire in the secondary coil. Owing to the rapidity of vibration of the primary currents, comparatively few turns of thick wire give the required pressure in the secondary circuit. To illustrate this feature by a practical experiment I take a simple paper cylinder, wound with only one layer of ordinary magnet wire, forming the secondary coil. In spite of there being only a few turns, long sparks—several inches in length—are obtained when the coil is inserted within or brought near to the discharge circuit of the instrument. A secondary of this form is simplest and best suitable for the production of long sparks, but it is somewhat inconvenient to handle.

The most advantageous features of these instruments lie, however, in the quality of the effects produced, which are the result of the rapidity or suddenness of the discharges obtained. To appreciate this feature we only need consider that a spark of, for instance, 6 inches in length, obtained with an instrument giving half a million vibrations a second, involves maximum pressures which, if produced with ordinary methods, would give sparks of many hundred feet, since the electrical force necessary to vibrate a certain quantity of electricity increases very rapidly; that is, with the square of the frequency of vibration. Therefore, pressures such as these here obtainable cannot be secured in any way by static machines or ordinary induction coils.

Still another feature of a more practical bearing I may illustrate by lighting a vacuum tube from an instrument furnishing currents of a frequency of much over half a million a second. Although the tube has a volume of only about 2½ [cubic] inches, it emits more light than a tube 6 or 7 feet long and 1½ inches in diameter, such as I have shown on other occasions, and that is a tube having 60 times the bulk and taking a proportionately larger amount of energy. So small a tube as this shown could not at all be brought to this luminosity by the use of the ordinary currents without soon getting overheated, and no better test of the increased efficiency of the light production can be had than by producing as high a luminosity in a small tube without undue heating.

Another convenient and advantageous feature of such an instrument will be found in its capability of being operated from alternating as well as from direct-current municipal circuits. With the special object in view of enabling their being used to the best advantage on alternating circuits also, I have determined the physical constants in a few types to suit circuits of the frequencies usually adopted here; that is, 60 or 125 cycles per second.

In the development and practical application of the principle underlying this kind of apparatus, one of the greatest difficulties encountered was the insulation of the secondary coils and condensers, particularly of the latter. The stored energy of a condenser is of an explosive nature, and when released suddenly in a way as it is in these instruments, it partakes much of the character of explosions of such a body as dynamite and enormous maximum pressures result, which strain the dielectric layers in the condensers and coils to their utmost. No matter how good and thick an insulation is provided, it cannot withstand such strains if there be even a slight absorption loss caused in any strained portion of the apparatus. An ordinary condenser, insulated as usual by thick layers of mica, which easily stands a few thousand volts of steady or slowly varying pressure, breaks down invariably; and no wonder it does; for, with vibrations of several hundred thousand a second, such a condenser with air bubbles or cavities of any kind, unavoidable when the usual method of construction is followed, will convert into heat the larger portion of the energy supplied to it. To investigate the flow of an alternating current through a coil with an iron core which is not laminated is hardly less crude than to carry on a research of rapid electrical vibrations with a condenser in which there are cavities or air bubbles, or in which, in general, air has access to the highly charged conductors. No estimate of the vibration period of an electromagnetic sys–tem can in such a case be made with any accuracy, whereas, when a proper plan of construction is followed and the dissipation of energy obviated, the experimental result closely agrees with the calculated period. By properly building up

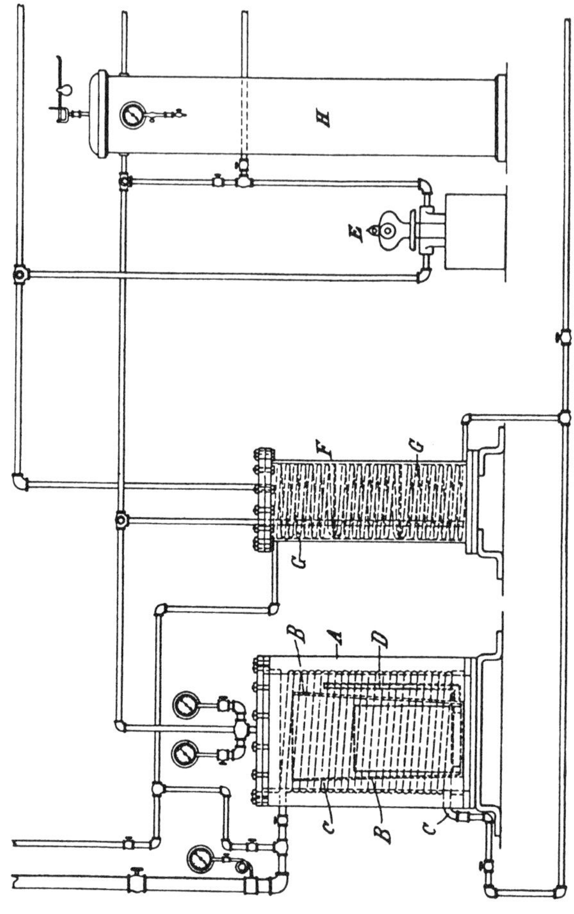

Fig. 10.—Apparatus for the manufacture of condensers and coils.

the condensers and coils, I have produced electromagnetic systems in which a slow vibration, once started, continues a minute or more, this indicating the absence of any serious friction loss. It is important to consider the preceding facts when dealing with standards and instruments of measure. A standard condenser prepared in the ordinary way of mica sheets and tinfoil, while indicating the correct value of capacity when used with a steady or slowly varying potential, will have its measured capacity greatly increased when the variation of potential becomes extremely rapid. In like manner, an electrostatic voltmeter with its vanes immersed in air, though a precious instrument with ordinary currents, is practically useless in the measurement of condenser discharges of frequencies of a few hundred thousand a second, its indication being far too low.

In view of the importance of the subject, a few words on the process of insulating, which has been adopted by me after several years of experimentation, may be of value. One form of apparatus as used by me is illustrated in diagram Fig. 10. *A* is a tank capable of withstanding great pressure, which is connected to a pump *E* and its reservoir *H* through a condensing reservoir *F*, kept cool by means of the coiled pipe *G*. The tank *A* is likewise provided with a coiled pipe *C*, through which either steam or cold water may be passed at will. The condenser is build up of insulating and conducting sheets in any convenient way, several layers of very thin paper being put together so as to avoid defects which may arise from small holes or punctures. For the same reason, it is advisable to mix up the sheets when received from the factory, as a great number of them may be injured at the same place. The condenser, having been tested by the application of moderate electrical pressure as that of a supply circuit of 220 volts, is placed in a tapering vessel *B*. A pipe *D*, reaching to the bottom of this vessel, may be provided, through which the insulation, when liquefied by the heat, may flow in, but this is of less importance. The vessel *B* containing the condenser being next placed in the tank *A*, and the top of the latter bolted down, steam is then passed through the coiled pipe *C* and the insulating mass is kept at the [proper] temperature which is a little above the melting point of the compound by regulating the steam supply. The pump is now connected with the tank by opening the proper valves, and a vacuum of about 29 inches or slightly more is established. When the melted compound has thoroughly permeated the interstices of the condenser, steam is then shut off and cold water passed through the coil *C*. The process of slow cooling being pushed far enough, the connections of the pump are reversed and air is forced into the tank *A* with the result of compressing strongly the fluid insulation and forcing it into all interstices. The pressure is preferably maintained until the mass is solidified. The application of the pressure is not only of great advantage because the insulation is forced into the interstices and prevented from shrinking away when cooling, but, in addition, any small gas bubble, which might remain in the condenser and would otherwise at ordinary atmospheric or smaller pressure be fatal to the instrument, is strongly compressed and the danger considerably lessened. The mass in the tank *A* being solidified, steam is again turned on the pipe *C* for a few minutes in order to soften the insulation on the periphery and allow the

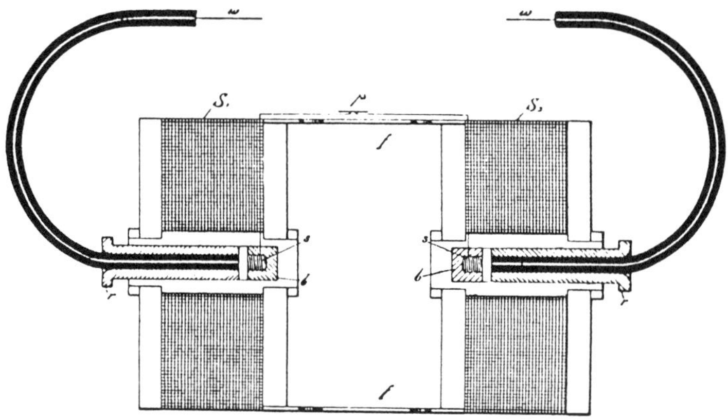

Fig. 11.—High potential coil system having terminals at centers.

vessel *B* to be lifted out of the tank, whereupon the condenser is taken out of the vessel and the superfluous insulation cut off. In the same manner, primary and secondary coils are treated. As insulating material, I have found best to use a mixture of beeswax and paraffin of low melting point, about half of each being taken. This gives a tough mass which [but slightly] shrinks away from the metal upon cooling. Condensers and coils manufactured in this manner will withstand incredible pressures. Very often in adjusting the primary discharge circuit, it may happen that sparks of 3/8 or 1/2 inch dart across the condenser terminals, and yet it will not break down, although the dielectric is no more than a few thousandth of an inch in thickness. I have been unable to detect any increase of temperature whatever in the condenser after long working.

To enable the secondary coils to withstand the effect of the enormous pressures producible with these instruments, I have recognized it as necessary to build them on the general plan illustrated in Fig. 11. The diagram shows two flat spirally wound coils, S_1 S_2, which are connected with their outer ends to a contact plate *p* in the proper direction so as to form in reality one single secondary coil, the terminals of which are respectively at the centers of the two wooden spools upon which the two parts of the coil are wound. These spools are held together by a cylinder of thin fiber sheet *ff*, which is sufficiently strong to insure solidity and perforated in order to allow the melted wax to fill the hollow spaces when the coil is put through the insulating process before described. In the centers of the spools are fastened threaded brass bushings *bb* to which the free ends of the secondary coils S_1 S_2 are connected and into which can be screwed brass pieces *ss*. The latter are fastened to the end of the hollow plugs of hard rubber *rr*, through which pass flexible wires *ww*, very heavily insulated with gutta-percha, which serve to connect the secondary high potential ends to the discharge rods supported on the top of the instrument (Fig. 9). It is advisable not to insulate the wires *ww* with soft rubber, for this kind of insulation is soon destroyed by the ozone generated at their surface in consequence of the streamers which will form even if the rubber be very thick. The thickness of the insulation between the superimposed layers of secondaries is practically determined from an approximate estimate of the difference of potential between adjacent layers. Originally I have used heavily insulated wires with from two to four braids, but presently I am using ordinary magnet wire which, in manufacturing the coil, is wound together with a string of a thickness equal to that of the wire. This is a convenient mode of insulating, not requiring specially prepared wire and secures excellent results. The middle of the secondary circuit, or common joint of the two coils, is connected to ground, or so the mains, and this generally through the primary discharge circuit. The small contact plate, or spring *p*, serves to establish the connection upon the secondary spools being inserted in the primary coil. The length of each of the secondary coils is so determined that it is somewhat less or equal to a quarter of the wavelength of the electromagnetic disturbance produced in the secondary circuit based, of course, on the practical estimate of the speed of propagation of the disturbance through this circuit. It is obviously understood that the length of the secondary circuit is made to approximate more or less a quarter of the wavelength, according to how much allowance is made for the capacity of the circuit under normal working conditions. In the ordinary uses of the instrument, as [a] laboratory appliance chiefly for the production of qualitative effects of high tension discharges, little allowance is generally made for the capacity of the terminals; but if the apparatus is designed, for instance for generating a large quantity of streamers between plates of great surface, or for charging condensers from the secondary, or [other] such uses, then the length of the secondary wire is made much smaller, and advantageously an even fraction of a quarter of that wavelength, which is obtained without any allowance for capacity other than that possessed by the coil. Finally, if secondary currents of comparatively low tension are desired, the coil is constructed preferably of one single spool and of only few layers, all in close proximity to the primary so as to increase the mutual induction coefficient and reduce the resonant rise of potential as much as possible. The closure of the magnetic circuit by oxygen at ordinary or high pressure, while of little effect with low frequency currents, is of a remarkable influence with currents of these unusually high frequencies, especially when the conditions are favorable for the occurrence of resonant phenomena, and I am anticipating practical uses of oxygen in this connection.

A secondary coil constructed in the manner illustrated in Fig. 11 has many important advantages, the chief ones being the safety in handling and the facility it affords for obtaining potentials far beyond those producible if the ordinary methods of construction are followed. In order to convey an idea of the pressures obtainable even with so small an instrument as the one described, a photograph of the same in action with two loops of cotton-covered wire attached to the discharge rods, is added (Fig. 12). The outer wire loop was in the experiment only 22 inches in diameter to enable it being properly shown in the print, but it could have been much larger since two such parallel wires 15 feet long may be stretched from the secondary terminals of the instrument and practically the whole space between them, 4 inches wide, is seen in the dark covered with fine luminous streamers. This is a surface of 5 square feet, and yet the energy taken from the supply circuit during the performance is less than 35 watts. To produce with an ordinary transformer such a quantity of these streamers, which may be needed for the manufacture of ozone or similar purposes, would require a considerably greater amount of energy and a more costly apparatus.

These extreme differences of potential obtainable by the use of the principle here involved are the result of the enormous suddenness or rate of change of the primary current impulses. In the ordinary method of varying the strength of the primary current, either by alternating the same or breaking the conducting path, we are limited to the comparatively insignificant rate of change producible by means of a high frequency alternator or rapid break, but by the use of the condenser discharges there is practically no limit to the suddenness of the impulses, and any potentials and spark lengths desired can be readily obtained. So, for instance, I have been able to produce, by applying the principle in a peculiar manner, immense electrical pressures, the theoretical maximum value of which can be measured only in many millions of volts, causing showers of continuous streams of thick, thundering sparks to dart out into space to a distance of 8 or 9 feet from an insulated wire, which behave sometimes like veritable lightening bolts and have afforded to the few who have witnessed them during the last two or three years in my laboratory a spectacle not easily forgotten. Nor is it at all difficult to increase, in a large hall or open space,

Fig. 12.—Photograph of coil system illustrated in figure 11 in action. Luminous streams cover an area of 5 square feet.

many times the potential and sparking distance by the employment of such means and methods.

Although in these oscillators the great suddenness of change in the strength of the currents depends chiefly on the electrical constants of the circuits, some advantages of minor but practical importance may be secured by a proper construction of the devices used as convenient, though not indispensable, accessories of the system for the purpose of arbitrarily making and breaking the circuits. Accordingly, I have devoted considerable time to their study and perfection, and in connection with the typical arrangements of the circuits illustrated in Figs. 1, 3, 4, and 5, I have dwelt in my earlier writings on this subject on a variety of such circuit interrupters in vacuum, air, and other fluids at low or great pressures.

It has been known long ago, since the investigations of Poggendorff, that, when the vibrator or break of an induction coil was enclosed in an exhausted vessel, the interruption of the currents was effected with greater suddenness, the vacuous space acting in a certain measure like a condenser, connected, as usual, around the break. My experiments with several kinds of such circuit breakers have led me to recognize that the vacuous space is not exactly the equivalent of a condenser, but rather of an absorbent, the increased suddenness being simply due to the rapid carrying away of the volatilized material forming the arc and, therefore, dependent on the velocity with which the disintegrated matter is carried away and also on the amount of the latter. Thus, with very hard platinum-iridium contacts and small currents, there is little difference; but, with soft platinum points and heavy currents, the influence of the vacuum is well noticeable, while, with mercury or in general easily volatilizable conductors, the difference is very great. The size of the exhausted vessel is also of some consequence, the break gaining in suddenness when the vessel is larger. Looking at Poggendorff's observations in this light, it appeared clear to me that only a small velocity of the particles composing the arc can be obtained since the effective pressure, at least with low frequency impulses depending on mechanical means, and with currents of limited strength which can be passed through the contacts without quickly destroying them, is necessarily only a minute fraction of the atmosphere being besides, very materially reduced by the oppositely acting attraction of the parallel-current elements of the arc. Pursuing further this train of reasoning, it seemed likewise evident that, if an insulating fluid be forced mechanically between the contact points with such velocity that the particles composing the arc were carried away quicker than it was possible with a small pressure producible in the gaseous matter in vacuum, the suddenness of disruption would be increased. This conclusion was borne out by my experiments in which I found that a fluid insulator, such as oil or alcohol, forced through the gap with even moderate velocity, increased very greatly the maximum rate of change of the primary current, and the length of secondary wire necessary for a certain spark length was in some instances reduced to 25 percent of that usually required. The length of the secondary was still further reduced by the use of insulating fluids under great pressure. As regards the suddenness of the current impulse following the closing of the contacts, the introduction of an insulating space or film of greater dielectric strength than that of the air at ordinary pressure, though producing a distinct effect, is of small consequence when the interrupter in its operation actually breaks the arc, since the electromotive force of a battery or municipal supply circuit is generally insufficient to break down an insulating film of even so small a thickness as 0.001 inch.

The continued effort to perfect the various automatic contrivances for controlling the supply current has clearly brought out their mechanical limitations, and the idea of utilizing the discharges of the condenser as a means for producing, independently of such mechanical devices, the sudden variations of the current which are needed for many purposes in the arts, appears evermore a happy and timely solution. In this novel process, a function of only minor importance is assigned to the mechanical means; namely, that of merely starting periodically the vibration of the electromagnetic system, and they have no other requirements to fulfill beyond those of reliability in operation and durability, features which are left to the skill of the mechanic and which, in a fair measure, it was not difficult to attain in a number of types.

Considering, then, that the rate of change of the discharge or primary current in these instruments is made to depend chiefly on the physical constants of the circuit through which the condenser discharges, it is evidently of utmost importance to construct properly the latter circuit, and in the investigations which were carried on with this object in view, several noteworthy observations have been made.

First of all, one draws the obvious conclusion that, inasmuch as the primary coil in a transformer of this kind consists usually of very few turns of copper ribbon of inappreciable resistances, the insulation between the turns should not require much care. But practical experience soon convinces him of his error, for, very often it happens that, owing to an exceptional resonant rise, the difference of potential between adjacent turns becomes so great as to rupture even a very good ordinary insulation. For this reason, it was found necessary to treat the primary coils likewise in the manner described, thus securing the additional advantage of stiffness, which results from the expansion of the metal sheets and thickening of the insulating layers during the heating in vacuum and subsequent contraction of the metal in cooling to the normal temperature after the insulation has solidified.

Next the experimenter is surprised when realizing the importance of the proper adjustment of the length of the primary coil and its connections. He is naturally prepared to find that, since the discharge circuit is of small length, the introduction in this circuit of a very small inductance or frictional resistance would produce an appreciable difference in the result obtained as, for instance, in the spark length of the secondary coil. But he certainly does not expect to observe that sometimes as little as 1/4 inch of conductor more or less would be of a telling effect. To illustrate: It is quite easy to produce with this kind of apparatus a spark of several feet in length, and by merely taking off or adding to the primary

1 inch of thick copper wire so reduce the spark length to one half. Observations of this kind impress the experimenter with the importance of the close adjustment of the circuits and accurate determination of their constants. His attention is forcibly attracted to the advantages of reducing as much as it is practicable the self-induction and resistance of the discharge circuit, the former with the object of securing the quickest possible vibration, the latter chiefly for reasons of economy. He also recognizes the necessity of bringing down to the minimum the length and resistance of all connecting wires. A well-constructed discharge circuit in a small instrument, such as the one described, should have no more than five percent of inactive conductor; its resistance should be negligible, and the self-induction should be not more than a few hundred centimeters.[*] I have found it almost imperative to use thin copper ribbon in the construction of the primary coils, and with these an observation, which is the most curious of all, has been made. It occurs, namely, that, under certain conditions, the primary coil gets perceptibly cooler by continued working. For a long time this result appeared doubtful, but finally it was positively ascertained and ascribed to an exaggerated Thomson effect, owing to which heat is carried from the primary copper ribbon to the tinfoil of the condenser.

It might not appear quite clear at first why the primary discharge circuit is so sensitive to variations of length, for a circuit of any length might be connected to the condenser and, provided that the relation between resistance, capacity and self-induction is such as to satisfy the condition laid down by Lord Kelvin, oscillatory discharge will take place. But it must be remembered that the velocity of propagation of the disturbance in the circuit depends on these quantities, and that the best result is attained when the velocity is such that a stationary wave is formed with a single node which is located generally, but not always, at a point of the discharge circuit or conductor equidistant from the opposite condenser coatings. Under such conditions, the maximum effective pressure at the terminals of the condenser is obtained. But

* In contemporary units, a few tenths microhenrys.

this state of things is only possible when the speed of the propagation through the discharge circuit is such that this circuit is traversed by the disturbance exactly in the time interval needed to complete half of one vibration. Now, since the speed is extreme and the length of the circuit very small, entirely insignificant variations of the length may often produce astonishing changes in the performance of the apparatus. These statements, of course, should not be construed as generally applicable, for they refer only to such cases in which the vibration in the discharge circuit, started by one operation of the circuit controller, does not die out before the succeeding operation of the controller. This may be made clear by a mechanical analogue. Suppose a weighted spring is clamped in a vise and a sudden blow is struck which sets the spring vibrating. Let the vibrations die out and let another blow be delivered. The spring will vibrate again as before, and it matters little what weight is attached to the spring, what the elasticity of the latter or, in general, what its period of vibration, and at what intervals the blows are delivered, the process of conversion of the energy of the blows into the energy of the vibrations will be effected with equal economy, except for secondary causes, immaterial for the present consideration. Exactly so is it with the electromagnetic system, and in the early stages of development and practical adaptation of the principle underlying the instruments described, I have employed condensers, either ordinary or electrolytic, of very large capacity and have caused them to discharge at comparatively long intervals through a primary circuit of negligible self-induction and resistance, thus producing current impulses which would sometimes reach, at least theoretically, maximum values of as much as 100,000 amperes. A high maximum rate of change in the primary current was thus producible, but, nevertheless, the average rate of change was still small. Considering again the mechanical analogue before mentioned, a valuable lesson is at once derived. Looking upon the weighted spring as an appliance for converting energy, both economy and output demand that the vibration of the spring should persist as long as possible and that the blows should be struck as often as it is practicable. To satisfy this twofold requirement, the blows must of necessity be delivered while the spring is still vibrating, and now it becomes most important to properly time the blows. Similarly again, in the electromagnetic system, the circuit controller must operate at definite intervals of time in order to secure the most vigorous vibration with the least supply of energy. In the construction of practical instruments, the number of the fundamental current impulses is arbitrarily adopted; the condenser, being prepared by a special process, cannot be adjusted without great inconvenience, and the size and to a certain extent also the turns of the primary coil are likewise determined beforehand from practical considerations. Furthermore, it is desirable, for reasons of economy, not to resort to an otherwise convenient method of adjustment, which would be to insert a variable self-induction in series with the primary coil. These conditions render more difficult the exact adjustment of the various quantities, and I have sometimes found it of advantage to adopt one or other plan such as will readily suggest themselves. For example, I have used an additional coil wound upon the primary and connected in parallel to the same, or I have completed the adjustments by determining properly the self-induction and capacity of the secondary coil.

In order to facilitate the observation and also to enable the exact determination of the oscillations of electromagnetic systems as well as of the vibrations or revolutions of mechanical devices, such as the circuit controllers used in connection, it was recognized as indispensable, in the course of these investigations, to construct a proper apparatus for such purposes. I determined from the outset to avail myself of what is known as visual synchronism. In this scheme, usually a disk or cylinder with marks or divisions, which is rotated with uniform velocity, is illuminated by a periodically varying or intermittent source of light, the divisions appearing stationary in space when the revolutions of the disk are synchronous with the variations in intensity or intermittence of the light-giving source. The chief virtue of such a method evidently resides in the uniformity of the velocity of rotation or eventually in the constancy of the period of the vibration produced. Having been early confronted with the problem of rotating a body with rigorously uniform velocity, which is required in many instances, or with the similar problem of producing a vibration of constant period, I have devoted some energy to the study of this subject, and in the course of time several solutions, more or less practical and satisfactory, have presented themselves.

One of these, for instance, was to produce by means of compressed air or steam, the vibration of a freely movable plunger to which was rigidly connected a coil or core of an electric generator. By the reciprocating motion of the plunger, alternating currents were generated which were passed through a condenser or else through the primary of a transformer, in which case the secondary coil of the latter was joined to the terminals of the condenser. Care being taken that the air or steam pressure was applied only during a short interval of time when the plunger was passing through the center of vibration, and the oscillations of the electromagnetic system, composed of the condenser and generating coil, being properly determined so that fundamental resonance took place, it was found that, under such conditions, the electromagnetic system entirely governed the vibrations of the plunger; the variations of the applied fluid pressure, while capable of producing changes in the amplitude, were within very wide limits without any appreciable effect on the period of vibration of the mechanical system, the currents generated being therefore of rigorously constant period. The currents thus obtained were then utilized in a number of ways to produce uniform rotation.

Another way to reach the same result and in a more practical manner was to generate currents of differing phase by a steam engine of special design, in which the reciprocating motion of the work performing plungers and attached magnetic cores or coils was controlled by a freely oscillating slide valve, the period of which was maintained constant by mechanical means or by the use of an electromagnetic system, similarly as before. A synchronous alternating motor operated by the two or three phase currents thus generated rotated with so uniform a velocity as to drive the wheel work of a clock with fair accuracy.

Still other solutions of the problems referred to I may mention which, though less satisfactory, have proved sometimes convenient and sufficient for many purposes. For example, a direct-current motor with laminated fields, or without any iron, was connected in series with a condenser through a commutator or interrupter fastened on the shaft of a light [weight] armature. This device was so constructed that it alternately closed and opened the terminals of the condenser as usual in the instruments before described. The condenser terminals being closed, a strong current impulse passed through the motor, and upon the terminals being opened the discharge current of high tension rushed into the condenser. But the energy and duration of both of these succeeding current impulses, and consequently of all which passed through the motor, were made chiefly dependent on the self-induction of the motor coils and on the capacity of the condenser and were, therefore, with certain limits of variation of the applied electromotive force, little dependent on the latter, and consequently a motor with a negligible friction loss, operated in this manner, turned with nearly uniform velocity. The latter was the more nearly constant the greater the controlling influence of the electromagnetic system which, of course, was the most complete when the number of current impulses, the capacity, and self-induction were so adjusted that fundamental resonance was maintained. As before stated, in most of these novel instruments described, such adjustments are observed and, whether provided with rotating interrupters or circuit-controlling springs, they partake more or less of the virtue of the preceding principle. For this reason, the contact springs in these instruments do not fall into harmonics, as they often do in ordinary induction coils operated from supply circuits where the physical constants are generally such that similar adjustments are impracticable.

It should be remarked that, since a long time, it was known that a direct-current motor, driven with currents interrupted at regular intervals, shows a marked tendency to maintaining a constant speed; but by the introduction of a condenser in the circuit and the careful adjustment of the quantities, this tendency is very much increased, and for many purposes a sufficiently uniform velocity may be obtained in this manner. Instead of using interrupted currents for operating the motor, it is practicable to rotate a separate coil, wound either on the same or on a second armature, and to pass the alternating currents generated in this coil through the condenser. It is important for the attainment of a satisfactory result in such cases to determine the constants so that the amount of energy stored in the condenser should be as large as possible.

While a number of such arrangements were readily available, it was found, nevertheless, that they were inadequate to the many different requirements of the laboratory, and accordingly an instrument was devised which is illustrated in Fig. 13 ab. It has proved itself to be so necessary and valuable an implement in experimental investigations that its description here may afford useful information. The cut is intended to show a substantial and carefully constructed clock mechanism with the usual escapement *e*, gearwheels *ggg*, and a 1-second pendulum *P*. A small shaft *s*, carrying a disk *D* of large diameter, was geared to the clockwork through a pinion *p* of a proper number of teeth, such as to give to the shaft a velocity best suitable for observations. Now, in order to rotate the disk with a uniform velocity, some difficulties, well known to clockmakers, had to be overcome. The chief of these is due to the fact that the rotation of the shaft *s*, being controlled by the escapement *e*, which, at regular intervals, retards the train of wheels *ggg*, is not effected with uniform but periodically varying velocity, which may have all values from zero to a maximum, dependent on the driving weight *W*. Owing to this circumstance, when such a disk *D* of large diameter is rigidly geared to any kind of clockwork, it exerts, by reason of the great momentum which it necessarily acquires, a strong reaction upon the pendulum, altering the period of the same more or less, according to the momentum it possesses. This difficulty is known to exist, even in cases in which the step by step movement is practically done away with, as, for instance, in clockworks with centrifugal governors, or circular pendulums, in which slow oscillations are produced by the reaction of the moving mass upon the regulating mechanism.

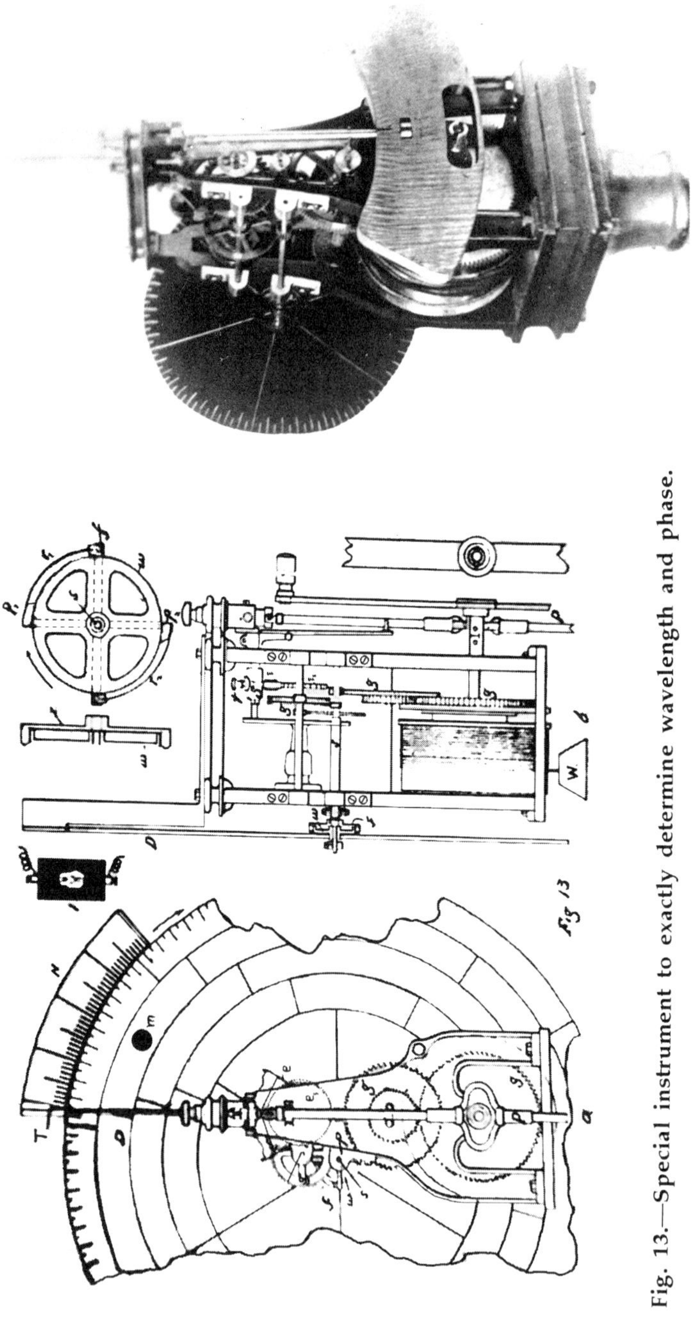

Fig. 13.—Special instrument to exactly determine wavelength and phase.

Some clockmakers have proposed an elastic connection between the body driven and the escapement, but this does not [do] away radically with the difficulty. On the other hand when, in an attempt to overcome this disadvantage of the step-by-step movement, a quick-acting escapement is used, whereby the periods of rest are reduced, and consequently the influence of the momentum of the rotated body upon the period of the pendulum, the result aimed at is but imperfectly attained and, besides, such an apparatus is less suitable for observation. Namely, it will be recognized as desirable for a number of reasons that the disk *D* should be rotated normally either once or twice a second, according to whether a 1- or ½-second pendulum is used. This being the case, the experimenter can render himself easily an account of the constancy of the speed by observing a mark *m* on the disk and noting that it occupies a fixed position in space, relative to that of the pendulum, in a convenient phase of vibration. Furthermore, the computation of the vibrations is rendered simpler and more convenient under such conditions.

The problem, clearly put, was then to rotate a body as the disk *D*, or other body, with any desired but uniform velocity in a way such that the period of vibration of the pen dulum was not much affected, even though the body rotated possessed considerable momentum. An entirely satisfactory solution of this problem was arrived at in the following manner. On the end of the shaft *s*, Fig. 13 b, was fastened a light metal piece *f* in the shape of a cross, carrying on two of its opposite sides pivoted pawls p_1 p_2, and on the other two light steel springs r_1 r_2, which pressed the pawls gently against the periphery of a washer *w*, which was provided with many very fine teeth or serrations cut sideways, similarly to those of escapement wheels. The washer *w* was arranged to turn very freely on the shaft *s*, and to it was fastened the disk *D*. The pawls p_1 p_2 were made with sharp edges to fit in the serrations of the washer *w*, and by these means the disk could rotate freely on the shaft *s* in the direction indicated by the arrows, but its rotation in the opposite direction was prevented by the pawls.

The operation of the apparatus will now be at once understood. On the start, the escapement wheel e was released by unscrewing the thumb screw t and shifting the sleeve S on its rocking support. The pendulum was next started and, when the escapement wheel had attained the normal velocity, the sleeve S was slipped back quickly and fastened—control of the escapement wheel being thus given to the pendulum. The wheel work and also the shaft s now moved with periodically varying velocity, but the disk D continued to move uniformly, the pawls p_1 p_2 slipping on the periphery of the washer w during the periods when the revolution of the shaft s was retarded by the pendulum. When, however, after some time, owing to the very small but unavoidable friction loss in the air and bearings, the speed of the disk would slowly diminish and fall below the maximum velocity which the shaft s was capable of imparting to it, then the pawls would give it a slight impulse, and in this manner the disk was kept constantly at the maximum velocity. By each swing of the pendulum, the disk would thus receive one impulse, and its velocity depended on the amount of energy imparted to it by each of the succeeding impulses. This amount of energy depended, of course, on the velocity of the shaft s during the period when the escapement wheel was free, and since this velocity was determined by the driving weight, the speed of the rotation of the disk could be varied within certain limits by adjusting the weight. It will be observed that, generally, the disk would rotate considerably faster than the shaft s, but it was easy to adjust the driving weight so that the disk rotated just once by one swing of the pendulum. In producing the rotation by these means, the influence of the momentum of the disk upon the period of the pendulum is found negligible. This result, of course, could not be attained by connecting the disk rigidly with the shaft s, even if a quick acting escapement would be used, as before suggested. The uniformity of rotation secured in this way leaves, for all practical purposes at least, nothing to be desired. The apparatus might have been improved by supporting the disk on an independent bearing and, perhaps, also by rotating it horizontally in a jeweled support. But the friction loss was very small, since, by arresting the shaft *s* suddenly, the disk would generally rotate something like 100 times or more before stopping, and such improvements were thought unnecessary. The vertical position was, however, chosen because it was much more convenient for purposes of observation. In order to reduce the weight of the disk *D* as much as possible, a light frame, consisting of a circular rim with narrow spokes, was cut out of thin aluminum sheet, and black paper glued on the frame—all marks and divisions of the former being, of course, white. I found it convenient to draw concentric circles with a number of marks such that all vibrations within the range of the apparatus could be read off. In addition, a segmental piece of hard rubber *N*, supported on a bar *T* and properly marked, was used to read fractions or, respectively, take corrections for any irregularity in the rotation during a prolonged period of time. Near the disk was placed a vacuum tube or, in its place, an adjustable spark gap *l*, which was connected to the secondary of a small transformer, the primary of which was positively controlled by the mechanical or electromagnetic system the vibrations of which were to be determined. In preparing a spring of the desired period of vibration for one of the instruments described, for instance, the spring was provisorily mounted on the instrument and the latter put in operation. The disk, intermittently illuminated by the discharges of the secondary coil, was released from the pendulum and rotated until synchronism was attained, the revolutions being computed by observing the white mark *m*. The constants of the spring were then modified after a simple calculation from the first result, and in the second trial, as a rule, the vibration was so close as to enable the use of the escapement, the adjustment then being completed, generally by altering the weight of the hammer on the spring until the marks on the disk, by the normal speed of rotation, appeared stationary in space.

The apparatus described in Fig. 13 will be found very convenient and time saving in a great many lines of experimentation. By means of the same, it is practicable to rotate a body of considerable weight with uniform and adjustable velocity, and it lends itself to the operation of circuit controllers, curve tracers, and all kinds of such devices. It will be found most useful in tracing current or electromotive-force curves and a variety of diagrams, and will afford material help in determining a number of physical quantities. But its most valuable use in the investigation of electrical vibrations is, perhaps, for the purpose of determining exactly the angular velocities of dynamos, particularly of alternators. Among the various quantities which, in alternate-current experimentation and practice, one has to determine very frequently, there are some, which even in a laboratory or shop in the midst of the disturbances of a city or factory can be ascertained with sufficient precision, while there are others which can be only approximated, particularly if, as is very often the case, practical methods of measure must be resorted to. So, for example, the close measurement of resistances offers no difficulty, nor does that of currents and electromotive forces, although the degree of exactitude is necessarily smaller; but in determining capacities, one is liable to make a considerable error, still a greater one in measuring inductances, and probably the greatest in estimating frequencies. In many places, such crude devices as speed counters or tachometers are still resorted to, and the experimenter is disappointed to realize that the accuracy of his long and painstaking tests is impaired because of his inability to determine exactly the frequency. To make matters worse, very often too, the latter is the largest and most important quantity. In view of these facts, a description of the method adopted by me for the determination of angular velocities may be of some value.

The devices commonly used are illustrated diagrammatically in Fig. 14, *a* and *b*. On the shaft *S*, Fig. 14 a, of the generator is fastened a commutator or circuit controller *C*, provided with any suitable number of segments, eight being

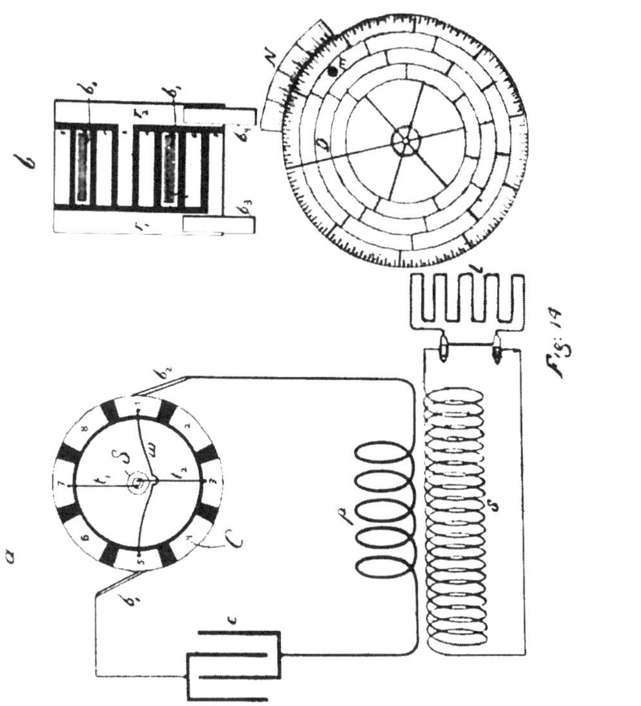

Fig. 14.—Method of impulse illumination of instrument disk.

shown in this instance. Four of these, *1, 3, 5,* and *7* serve to establish the connections of the circuits, while the intermediate ones, *2, 4, 6,* and *8* are entirely insulated, idle segments. Assuming the generator to be an alternate-current machine, the terminals t_1 t_2 of the armature winding, or of any desired coil or part of the same, are led through the hollow shaft, as may be the case, and connected to the diametrically opposite segments *3* and *7*, while the segments situated at right angles, that is *1* and *5*, are connected together through a wire w of inappreciable resistance. Two brushes b_1 b_2, supported in an ordinary holder allowing their being shifted in any position, are arranged to bear upon the periphery of the controller *C*. These brushes are connected to a circuit comprising a condenser *c* of proper capacity and a primary coil *p*, which has but a few turns of very small self-induction and resistance and is joined in series with the condenser.

The operation of the devices is as in the instruments before referred to. When, with the rotation of the shaft *S*, the brushes b_1 b_2 are brought in contact with the segments *1* and *3*, the condenser is charged to a potential which can be adjusted at will by shifting the brush holder. The condenser retains a certain charge until the brushes b_1 b_2 come to bear upon the connected segments *1* and *5*, whereupon an oscillatory discharge through the primary *p* takes place with the result of inducing strong current impulses in the secondary *s*, which momentarily light up the vacuum tube or spark gap *l* placed in proximity of the disk *D*, which is rotated with uniform velocity, as before described. With the rotation of the circuit controller, the brushes are again brought in contact with the segments *1* and *3*, and the operations are repeated, at each complete revolution of the armature shaft a definite number of impulses being passed through the vacuum tube or spark gap. In the device illustrated, there will be only two impulses for each revolution of the armature, but any greater number may be arranged for by augmenting the number of the segments and connecting them in the same manner. It should be stated that the current impulses, which pass into the condenser whenever the brushes b_1 b_2 are on those segments which are connected to the armature coil, ordinarily produce no appreciable effect in the secondary *s*. This might be the case if the number of segments would be very large and would then be at once noted. The proper adjustment of the circuit through which the condenser discharges is, of course, preferable but not absolutely necessary.

When it is inconvenient to use the armature current, as illustrated in Fig. 14 a, then the controller *C* is provided with two sliding rings r_1 r_2, Fig. 14 b, upon which are made to bear two additional brushes b_3 b_4. The latter are then connected to a direct-current source as the ordinary supply circuit, preferably through a self-induction coil, which serves to charge the condenser to a higher potential. The rings r_1 r_2 merely convey to the segments *1* and *3* the current for charging the condenser, otherwise nothing need be changed on the devices.

The marks or divisions on the periphery of the disk *D* are suitably made so that by the normal speed of the generator they appear stationary in space. This being the case, the speed may be at once and easily computed from the number of segments on the controller and that of divisions on the disk and from the speed of the latter. The frequency of the dynamo currents is then given by taking into consideration the number of poles.

In availing himself of this method, the experimenter can get the accurate value for the angular velocity, no matter how much the speed of the dynamo may vary, if he only takes the precaution to make his readings for electromotive force, current, etc., at the instant the marks on the disk are stationary. Should the reading consume more time, it is easy to take the correction for any variation by simply observing, with reference to a fixed line on the rubber piece *N*, the number of divisions which are to be added to, or deducted from, the speed of the disk.

Section I - Addendum

Wireless Telegraphy Receiving Methods; Advances in Electrical Oscillators; Experimental Vacuum Bulbs.

This portion of the lecture was not completed for publication by Tesla. The apparatus and drawings presented in the demonstration lecture on the subject of wireless telegraphy receiving methods, an extension of his presentation on novel high frequency measurement methods, were considered by him too revealing in terms of patent applications in progress. The following Addendum section is derived from Section IX, "Arrangements for receiving," *Nikola Tesla On His Work With Alternating Currents*, and is believed to summarize his remarks on this subject.

"The construction [of the instrument shown in Section I, Fig. 13 cut] was intended to produce an absolutely constant rotation so that certain intervals of time could be definitely fixed, and in relation to these intervals of time I could analyze the waves... The bottom of the diagram [Fig. 14] shows vacuum tubes designed for very minute currents. They were excited by the secondary of the transformer and illuminated the dial. If I used, for instance, two vibrations of different wavelength, then there was a beat, and I would notice, as this disc rotated, the marked lines travel one way or the other. When perfect synchronism was obtained, these lines appeared stationary.

"I am now showing [Fig. 15, top] a [drawing of a] device for telephonic and telegraphic signals I have used in my laboratory on Houston Street... That [left] is a transmitter ..., and [below] is an inductance which is bridged by a device such as that by speaking into it, or actuating it by hand or otherwise, variations in the intensity of the waves are produced.

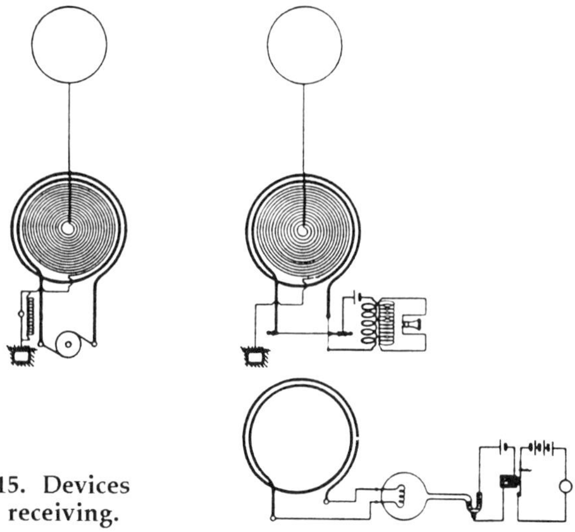

Fig. 15. Devices for receiving.

"On the receiver side [right] I have my antenna and self-inductance coil connected to the ground, and in the secondary I have a wire which is under a tension. Another wire, likewise under tension, controls two microphonic contacts or carbons. The tension of this wire is adjustable, and as I will show in another drawing, I can regulate the pressure of the contacts so that a certain current from a battery, here, will flow through this primary coil.

"When the transmitted oscillations are controlled and produce corresponding variations in the intensity of the received effects, then the current generated in [the secondary of the receiver] heats that wire more or less and the alternate heating and cooling of the latter results in periodic expansions and contractions vary[ing] the microphonic pressure of the contacts in obedience to the changes produced in the transmitter. In the secondary [of the transformer], I have a telephone [receiver] specially wound to reproduce the speech..."

"My transmitter was on Houston Street and I would take the receiver with me. For instance, I would take a few toy balloons, go on the roof, and then put my box there with the instruments and listen to the signals.

"This [Fig. 15, bottom] is another [drawing of a] device which I also used with success, but not telephonic. It operated on the principle of the Reis air thermometer ... [I]n the bulb is a resistance wire which is heated and cooled, owing to the fluctuations of the received currents. The attendant expansions and contractions of the air operate a little mercury column, pushing it back and forth. Curiously enough, for receiving telegraphic signals, this crude instrument was certainly good, but of course it was not suited for telephonic reception.

"That [shown in Fig. 16] ... illustrates a way of producing audible notes by reaction of the received impulses upon a magnetic field. [At upper left] is a transmitter, diagrammatically represented, with an arrangement for varying the intensity of the waves emitted, and on the receiver side I have, as you see, a grounded antenna. [The] secondary [has a conductor under tension in] a very powerful magnetic field, and [the reaction of] this conductor, traversed by the received currents in the field, causes the conductor to emit audible notes.

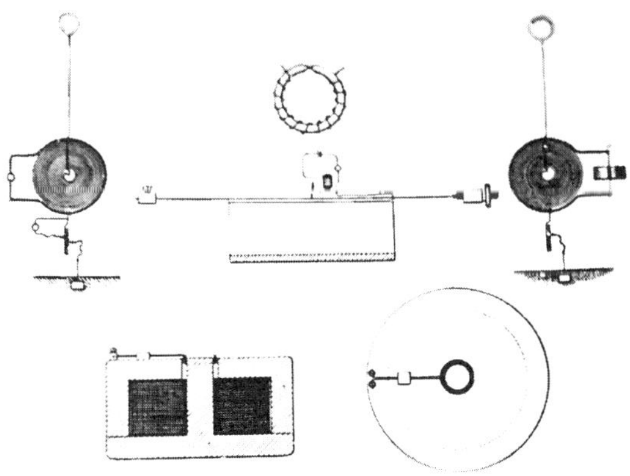

Fig. 16. Other Ways of receiving.

"I [have] several magnets of various forms, like this [Fig. 16, center], and employed a cord in the field, which, when the current traversed it, vibrated and established a contact. Or, I [use] a small coil ... through which the current was passed, and which by its vibrations produced the signal, an audible note, or anything else ... [I]n my writings ... I had already shown the reaction of the high frequency and low frequency currents on magnetic fields, and had specified the frequencies within which one has to keep in order to receive efficiently audible notes."

In addition to the electrical oscillator unit shown in Fig. 9, Section I, Tesla also exhibited two other units. The first is shown in Fig. 17 which was covered by a patent applied for nine months earlier.[49] A second is shown in Fig. 18 an advance look at a form of oscillator utilizing one of a series of eight hermetically-sealed, mercury circuit controllers for which patents were applied beginning the following two months. This unit was covered by a patent applied for eight months later.[50]

Fig. 17

Fig. 18

These units were described by Tesla the following way years later as presented for the lecture.

[49] U.S. Patent No. 568,179 of Sept. 22, 1896, "Method and Apparatus for Producing Currents of High Frequency," application filed July 6, 1896.

[50] U.S. Patent No. 609,245 of Aug. 16, 1898, "Electrical-Circuit Controller," application filed Dec. 2, 1897

"[The unit in Fig. 17] represents a large oscillator ... intended for wireless experiments, production of Roentgen rays, and scientific research in general. It comprises a box containing two condensers of the same capacity on which are supported the charging coil and transformer. The automatic circuit controller, hand switch and connecting posts are mounted on the front plate of the inductance spool as is also one of the contact springs. The condenser box is equipped with three terminals, the two external ones serving merely for connection while the middle one carries a contact bar with a screw for regulating the interval during which the circuit is closed. The vibrating spring itself, the sole function of which is to cause periodic interruptions, can be adjusted in its strength as well as distance from the iron core in the center of the charging coil by four screws visible on the top plate so that any desired conditions of mechanical control might be secured. The primary coil of the transformer is of copper sheet and taps are made at suitable points for the purpose of varying, at will, the number of turns. The inductance coil is wound in two sections to adapt the instrument both to 110 and 220 volt circuits and several secondaries were provided to suit the various wavelengths of the primary. The output was approximately 500 watts with damped waves about 50,000 cycles per second. For short periods of time undamped oscillations were produced in screwing the vibrating spring tight against the iron core and separating the contacts by the adjusting screw *which also performed the function of a key.*

"[The unit in Fig. 18] illustrates a transformer with a rotary break. There are two condensers of the same capacity in the box which can be connected in series or multiple. The charging inductances are in the form of two long spools upon which are supported the secondary terminals. A small direct-current motor, the speed of which can be varied within wide limits, is employed to drive a specially constructed make and break. In other features, the oscillator is like the one illustrated [at left] and its operation will be readily understood from the foregoing. This transformer was used in my wireless experiments and frequently also for lighting the laboratory by my vacuum tubes."[51]

[51] Tesla, "Electrical Oscillators," *Electrical Experimenter,* July 1919, pp. 228-229, 259-260, 276, 276.

"[I now show on the wall of this Academy drawings of] a great variety of bulbs I used. Every one that you see was built, not in one, but in several forms ... Among these bulbs I have a great number of receiving devices ...."

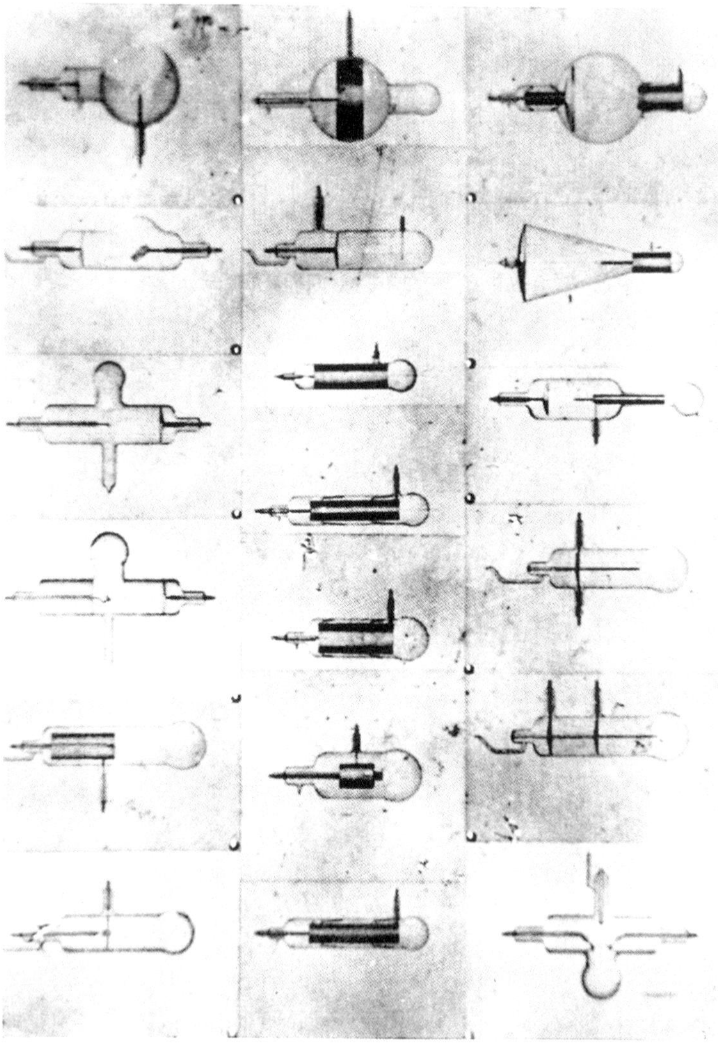

Group I

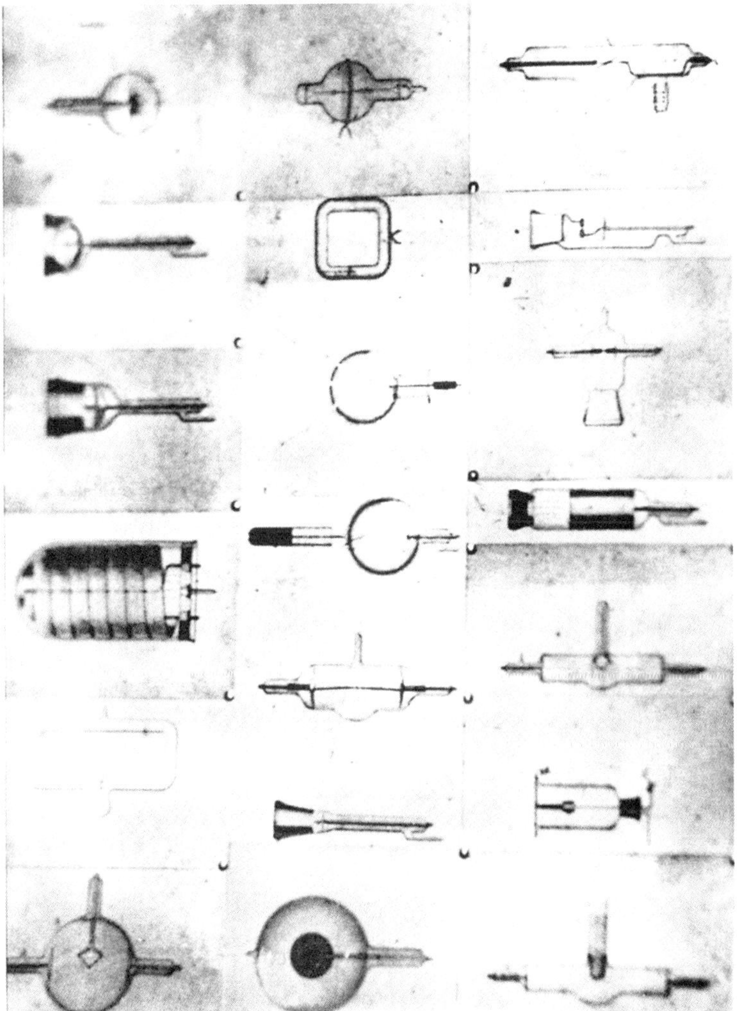

Group II

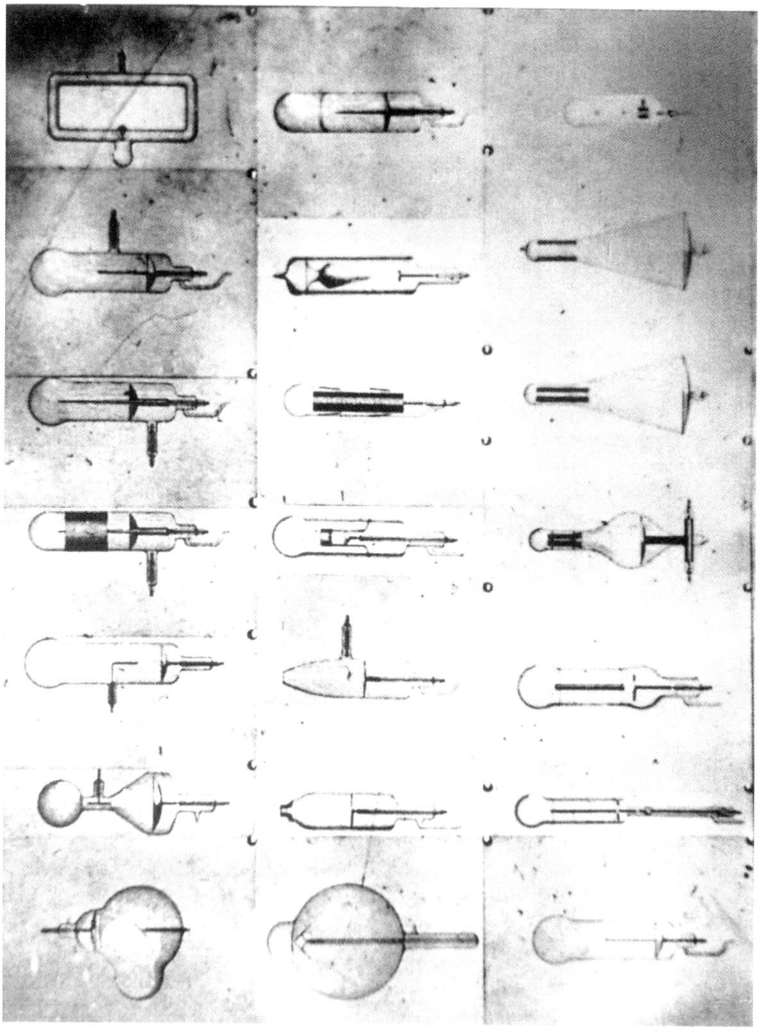

Group III

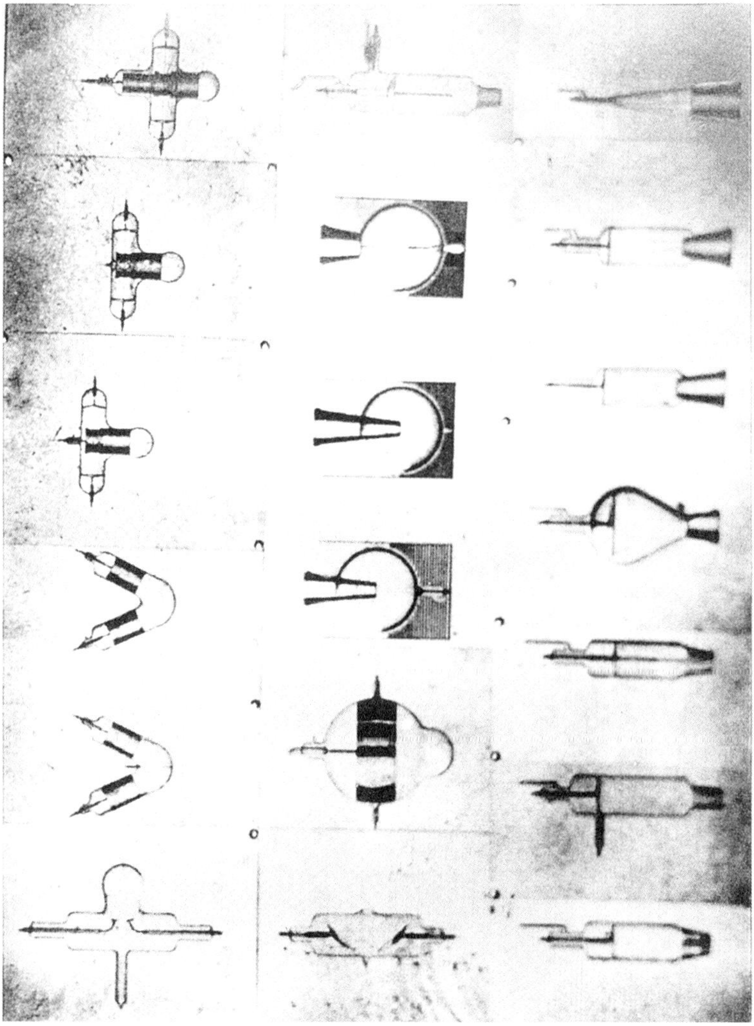

Group IV

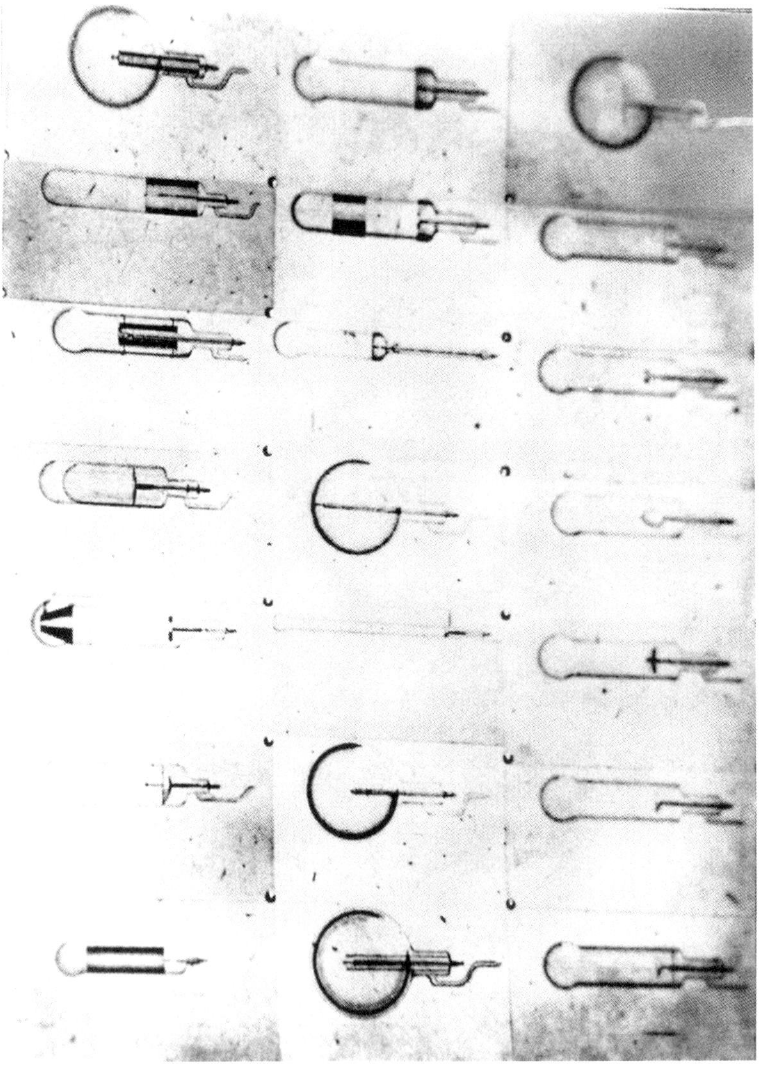

Group V

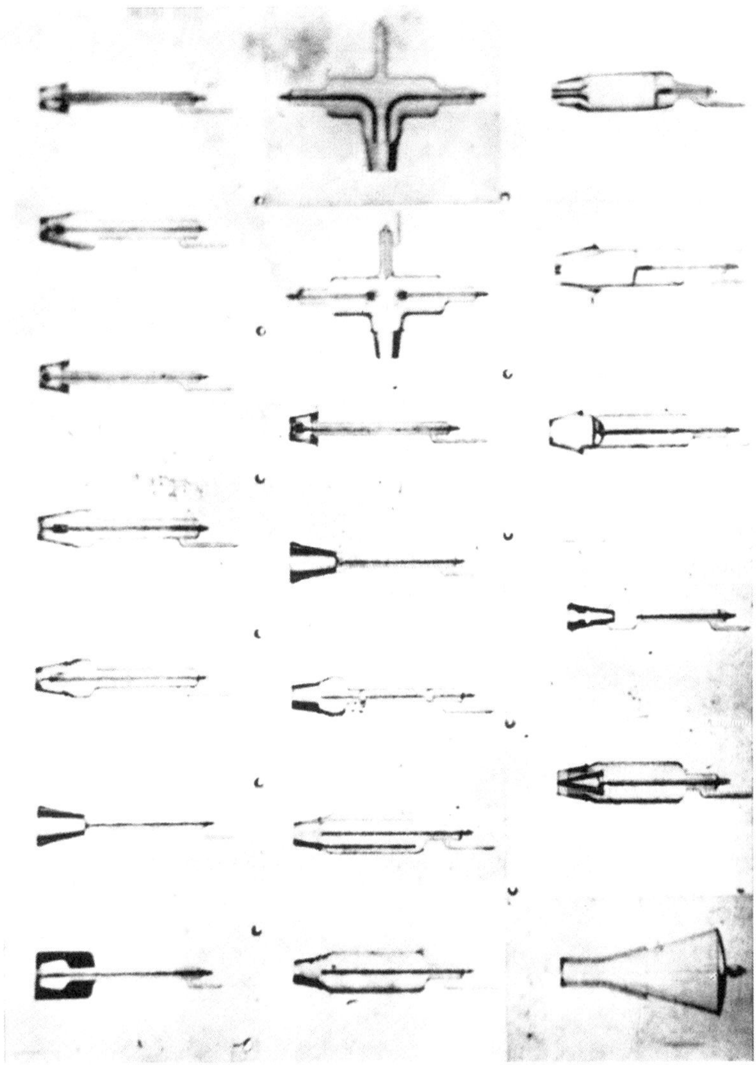

Group VI

Section II

THE HURTFUL ACTIONS OF LENARD AND ROENTGEN TUBES.

To the Editor of Electrical Review:

The rapidly extending use of the Lenard and Roentgen tubes or Crookes bulbs as implements of the physician, or as instruments of research in laboratories, makes it desirable, particularly in view of the possibility of certain hurtful actions on the human tissues, to investigate the nature of these influences, to ascertain the conditions under which they are liable to occur and --what is most important for the practitioner—to render all injury impossible by the observance of certain rules and the employment of unfailing remedies.

As I have stated in a previous communication to your esteemed journal (see *Electrical Review* of December 2, 1896), no experimenter need be deterred from using freely the Roentgen rays for fear of a poisonous or deleterious action, and it is entirely wrong to give room to expressions of a kind such as may tend to impede the progress and create a prejudice against an already highly beneficial and still more promising discovery; but it cannot be denied that it is equally uncommendable to ignore dangers now when we know that, under certain circumstances, they actually exist. I consider it the more necessary to be aware of these dangers, as I foresee the coming into general use of novel apparatus, capable of developing rays of incomparable great power. In scientific laboratories the instruments are usually in the hands of persons skilled in their manipulation and capable of approximately estimating the magnitude of the effects, and the omission of necessary precautions is, in the present state of our knowledge, not so much to be apprehended; but the physicians, who are keenly appreciating the immense benefits derived from the proper application of the new principle, and the numerous amateurs who are fascinated by the beauty of the novel manifestations, who are all passionately bent upon experimentation in the newly opened up fields, but many of whom are naturally not armed with the special knowledge of the electrician—all of these are much in need of reliable information from experts, and for these chiefly the following lines are written. However, in view of the still incomplete knowledge of these rays, I wish the statements which follow to be considered as devoid of authoritativeness, other than that which is based on the conscientiousness of my study and the faith in the precision of my senses and observations.

Ever since Professor Roentgen's discovery was made known I have carried on investigations in the direction indicated by him, and with perfected apparatus, producing rays of much greater intensity than it was possible to obtain with the usual appliances. Commonly, my bulbs were capable of showing the shadow of a hand on a phosphorescent screen at distances of 40 or 50 feet, or even more, and to the action of these bulbs myself and several of my assistants were exposed for hours at a time, and although the exposures took place every day, not the faintest hurtful action was noted —as long as certain precautions were taken. On the contrary, be it a coincidence, or an effect of the rays, or the result of some secondary cause present in the operation of the bulbs—as, for example, the generation of ozone—my own health, and that of two persons who were daily under the influence of the rays, more or less, has materially improved, and, whatever be the reason, it is a fact that a troublesome cough with which I was constantly afflicted has entirely disappeared, a similar improvement being observed on another person.

In getting the photographic impressions or studying the rays with a phosphorescent screen, I employed a plate of thin aluminum sheet or a gauze of aluminum wires, which was interposed between the bulb and the person, and connected to the ground directly or through a condenser. I adopted this precaution because it was known to me, a long time before, that a certain irritation of the skin is caused by very strong streamers, which, mostly at small distance, are formed on the body of a person through the electrostatic influence of a terminal of alternating high potential. I found that the occurrence of these streamers and their hurtful consequence was completely prevented by the employment of a conducting object, as a sheet of wire gauze placed and connected as described. It was observed, however, that the injurious effects mentioned did not seem to diminish gradually with the distance from the terminal, but ceased abruptly, and I could give no other explanation for the irritation of the skin which would be as plausible as that which I have expressed; namely, that the effect was due to ozone, which was abundantly produced. The latter peculiarity mentioned was also in agreement with this view, since the generation of ozone ceases abruptly at a definite distance from the terminal, making it evident that a certain intensity of action is absolutely required, as in a process of electrolytic decomposition.

In carrying further my investigations, I gradually modified the apparatus in several ways, and immediately I had opportunities to observe hurtful influences following the exposures. Inquiring now what changes I had introduced, I found that I had made three departures from the plan originally followed: First, the aluminum screen was not used; second, a bulb was employed which, instead of aluminum, contained platinum, either as electrode or impact plate; and third, the distances at which the exposures took place were smaller than usual.

It did not require a long time to ascertain that the interposed aluminum sheet was a very effective remedy against injury, for a hand could be exposed for a long time behind it without the skin being reddened, which otherwise invariably and very quickly occurred. This fact impressed me with the conviction that, whatever the nature of the hurtful influences, it was in a large measure dependent either on an electrostatic action, or electrification, or secondary effects resulting therefrom, such as are attendant to the formation of streamers. This view afforded an explanation why an observer could watch a bulb for any length of time, as long as he was holding the hand in front of the body, as in examining with a fluorescent screen, with perfect immunity to all parts of his body, with the exception of the hand. It likewise explained why burns were produced in some instances on the opposite side of the body, adjacent to the photographic plate, whereas portions on the directly exposed part of the body, which were much nearer to the bulb, and consequently subjected to by far stronger rays, remained unaffected. It also made it easy to understand why the patient experienced a prickling sensation on the exposed part of the body whenever an injurious action took place. Finally, this view agreed with the numerous observations that the hurtful actions occurred when air was present, clothing, however thick, affording no protection, while they practically ceased when a layer of a fluid, quite easily penetrated by the rays, but excluding all contact of the air with the skin, was used as a preventive.

Following, now, the second line of investigation, I compared bulbs containing aluminum only with those in which platinum was used besides, ordinarily as impact body, and soon there were enough evidences on hand to dispel all doubt as to the latter metal being by far the more injurious. In support of this statement, one of the experiences may be cited which, at the same time, may illustrate the necessity of taking proper precautions when operating bulbs of very high power. In order to carry out comparative tests, two tubes were constructed of an improved Lenard pattern, in size and most other respects alike. Both contained a concave cathode or reflector of nearly two inches in diameter, and both were provided with an aluminum cap or window. In one of the tubes the cathodic focus was made to coincide with the center of the cap, in the other the cathodic stream was concentrated upon a platinum wire supported on a glass stem axially with the tube a little in front of the window, and in each case the metal of the latter was thinned down in the central portion to such an extent as to be barely able to withstand the inward air pressure. In studying the action of the tubes, I exposed one hand to that containing aluminum only, and the other to the tube with the platinum wire. On turning on the former tube, I was surprised to observe that the aluminum window emitted a clear note, corresponding to the rhythmical impact of the cathodic stream. Placing the hand quite near the window, I felt distinctly that something warm was striking it. The sensation was unmistakable, and quite apart from the warmth felt, differed very much from that prickling feeling produced by streamers or minute sparks. Next I examined the tube with the platinum wire. No sound was emitted by the aluminum window, all the energy of the impact being seemingly spent on the platinum wire, which became incandescent, or else the matter composing the cathodic stream was so far disintegrated that the thin metal sheet offered no material obstruction to its passage. If big lumps are hurled against a wire netting with large meshes, there is considerable pressure exerted against the netting; if, on the contrary—for illustration—the lumps are very small as compared with the meshes, the pressure might not be manifest. But, although the window did not vibrate, I felt, nevertheless, again, and distinctly, that something was impinging against the hand, and the sensation of warmth was stronger than in the previous case. In the action on the screen there was apparently no difference between the two tubes, both rendering it very bright, and the definition of the shadows was the same, as far as it was possible to judge. I had looked through the screen at the second tube a few times only when something detracted my attention, and it was not until about 20 minutes later, when I observed that the hand exposed to it was much reddened and swollen. Thinking that it was due to some accidental injury, I turned again to the examination of the platinum tube, thrusting the same hand close to the window, and now I felt instantly a sensation of pain, which became more pronounced when the hand was placed repeatedly near the aluminum window. A peculiar feature was that the pain appeared to be seated, not at the surface, but deep in the tissues of the hand, or rather in the bones. Although the aggregate exposure was certainly not more than half a minute, I had to suffer severe pain for a few days afterward, and some time later I observed that all the hair was destroyed and that the nails on the injured hand had grown anew.

The bulb containing no platinum was now experimented with, more care being taken, but soon its comparative harmlessness was manifest, for, while it reddened the skin, the injury was not nearly as severe as with the other tube. The valuable experiences thus gained were: The evidence of something hot striking the exposed member; the pain *instantly* felt; the injury produced *immediately* after the exposure, and the increased violence due, in all probability, to the presence of the platinum.

Some time afterward I observed other remarkable actions at very small distances from powerful Lenard tubes. For instance, the hand being held near the window only for a few seconds, the skin seems to become tight, or else the muscles are stiffened, for some resistance is experienced in closing the fist, but upon opening and closing it repeatedly the sensation disappears, apparently no ill effect remaining. I have, furthermore, observed a decided influence on the nasal discharge organs similar to the effects of a cold just contracted. But the most interesting observation in this respect is the following: When such a powerful bulb is watched for some time, the head of the observer being brought very close, he soon after that experiences a sensation so peculiar that no one will fail to notice it when once his attention is called to it, it being almost as positive as touch. If one imagines himself looking at something like a cartridge, for instance, in close and dangerous proximity, and just about to explode, he will get a good idea of the sensation produced, only, in the case of the cartridge, one cannot render himself an account where the feeling exactly resides, for it seems to extend all over the body, this indicating that it comes from a general awareness of danger resulting from previous and manifold experiences, and not from the anticipation of an unpleasant impression directly upon one of the organs, as the eye or the ear; but, in the case of the Lenard bulb, one can at once, and with precision, locate the sensation: it is in the head. Now, this observation might not be of any value except, perhaps, in view of the peculiarity and acuteness of the feeling, were it not that exactly the same sensation is produced when working for some time with a noisy spark gap, or, in general, when exposing the ear to sharp noises or explosions. Since it seems impossible to imagine how the latter could cause such a sensation in any other way except by directly impressing the organs of hearing, I conclude that a Roentgen or Lenard tube, working in perfect silence as it may, nevertheless produces violent explosions or reports and concussions, which, though they are inaudible, take some material effect upon the bony structure of the head. Their inaudibility may be sufficiently explained by the well founded assumption that not the air, but some finer medium, is concerned in their propagation.

But it was in following up the third line of inquiry into the nature of these hurtful actions, namely, in studying the influence of distance, that the most important fact was unearthed. To illustrate it popularly, I will say that the Roentgen tube acts exactly like a source of intense heat. If one places the hand near to a red hot stove, he may be instantly injured. If he keeps the hand at a certain small distance, he may be able to withstand the rays for a few minutes or more, and may still be injured by prolonged exposure; but if he recedes only a little farther, where the heat is slightly less, he may withstand the heat in comfort and any length of time without receiving any injury, the radiations at that distance being too weak to seriously interfere with the life process of the skin. This is absolutely the way such a bulb acts. Beyond a certain distance, no hurtful effect whatever is produced on the skin, no matter *how long* the exposure. The character of the burns is also such as might be expected from a source of high heat. I have maintained, in all deference to the opinions of others, that those who have likened the effects on the skin and tissues to sunburns have misinterpreted them. There is no similarity in this respect, except insofar as the reddening and peeling of the skin is concerned, which may result from innumerable causes. The burns, when slight, rather resemble those people often receive when working close to a strong fire. But when the injury is severe, it is in all appearances like that received from contact with fire or from a red-hot iron. There may be no period of incubation at all, as is evident from the foregoing remarks, the rays taking effect immediately, not to say instantly. In a severe case the skin gets deeply colored and blackened in places, and ugly, ill-foreboding blisters form; thick layers come off, exposing the raw flesh, which, for a time, discharges freely. Burning pain, feverishness, and such symptoms are of course but natural accompaniments. One single injury of this kind, in the abdominal region, to a dear and zealous assistant—the only accident that ever happened to any one but myself in all my laboratory experience—I had the misfortune to witness. It occurred before all these and other experiences were gained, following directly an exposure of five minutes at the fairly safe distance of 11 inches to a very highly-charged platinum tube, the protecting aluminum screen having been unfortunately omitted, and it was such as to fill me with the gravest apprehensions. Fortunately, frequent warm baths, free application of Vaseline, cleaning, and general bodily care soon repaired the ravages of the destructive agent, and I breathed again freely. Had I known more than I did of these injurious actions, such unfortunate exposure would not have been made; had I known less than I did, it might have been made at a smaller distance, and a serious, perhaps irremediable, injury might have resulted.

I am using the first opportunity to comply with the bitter duty of recording the accident. I hope that others will do likewise, so that the most complete knowledge of these dangerous actions may soon be acquired. My apprehensions led me to consider, with keener interest than I would have felt otherwise, what the probabilities were in such a case of the internal tissues being seriously injured. I came to the very comforting conclusion that, no matter what the rays are ultimately recognized to be, practically all their destructive energy must spend itself on the surface of the body, the internal tissues being, in all probability, safe, unless the bulb would placed in very close proximity to the skin, or else, that rays of far greater intensity than now producible were generated. There are many reasons why this should be so, some of which will appear clear from my foregoing statements referring to the nature of the hurtful agencies, but I may be able to cite new facts in support of this view. A significant feature of the case reported may be mentioned. It was observed that on three places, which were covered by thick bone buttons, the skin was entirely unaffected, while it was entirely destroyed under each of the small holes in the buttons. Now, it was impossible for the rays, as investigation showed, to reach these points of the skin in straight lines drawn from the bulb, and this would seem to indicate that not all the injury was due to the rays or radiations under consideration, which unmistakably propagate in straight lines, but that, at least in part, concomitant causes were responsible. A further experimental demonstration of this fact may be obtained in the following manner: The experimenter may excite a bulb to a suitable and rather small degree, so as to illuminate the fluorescent screen to a certain intensity at a distance of, say 7 inches. He may expose his hand at that distance, and the skin will be reddened after a certain duration of exposure. He may now force the bulb up to a much higher power, until, at a distance of 14 inches, the screen is illuminated even stronger than it was before at half that distance. The rays are now evidently stronger at the greater distance, and yet he may expose the hand a very long time, and it is safe to assert that he will not be injured. Of course, it is possible to bring forth arguments which might deprive the above demonstration of force. So, it might be stated, that the actions on the screen or photographic plate do not give us an idea as to the density and other quantitative features of the rays, these actions being entirely of a qualitative character. Suppose the rays are formed by streams of material particles, as I believe, it is thinkable that it might be of no particular consequence, insofar as the visible impression on the screen or film is concerned, whether a trillion of particles per square millimeter strike the sensitive layer or only a million, for example; but with the actions on the skin it is different; these must surely and very materially depend on the quantity of the streams.

As soon as the before-mentioned fact was recognized, namely, that beyond a certain distance even the most powerful tubes are incapable of producing injurious action, no matter how long the exposure may last, it became important to ascertain the safe distance. Going over all my previous experiences, I found that, very frequently, I have had tubes which at a distance of 12 feet, for illustration, gave a strong impression of the chest of a person with an exposure of a few minutes, and many times persons have been subjected to the rays from these tubes at a distance of from 18 to 24 inches, the time of exposure varying from 10 to 45 minutes, and never the faintest trace of an injurious action was observed. With such tubes I have even made long exposures at distances of 14 inches, always, of course, through a thin sheet or wire gauze of aluminum connected to the ground, and, in each case, observing the precaution that the metal would not give any spark when the person was touching it with the hand, as it might sometimes be when the electrical vibration is of extremely high frequency, in which case a ground connection, through a condenser of proper capacity, should be resorted to. In all these instances bulbs containing only aluminum were used, and I therefore still lack sufficient data to form an exact idea of what distance would have been safe with a platinum tube. From the case previously cited, we see that a grave injury resulted at a distance of 11 inches, but I believe that, had the protecting screen been used, the injury, if any, would have been very slight. Taking all my experiences together, I am convinced that no serious injury can result if the distance is greater than 16 inches and the impression is taken in the manner I have described.

Having been successful in a number of lines of inquiry pertaining to this new department of science, I am able at present to form a broader view of the actions of the bulbs, which, I hope, will soon assume a quite definite shape. For the present, the following brief statement may be sufficient. According to the evidences I am obtaining, the bulb, when in action, is emitting a stream of small material particles. There are some experiments which seem to indicate that these particles start from the outer wall of the bulb; there are others which seem to prove that there is an actual penetration of the wall, and in the case of a thin aluminum window, I have now not the least doubt that some of the finely disintegrated cathodic matter is actually forced through. These streams may simply be projected to a great distance, the velocity gradually diminishing without the formation of any waves, or they may give rise to concussions and longitudinal waves. This, for the present consideration, is entirely immaterial, but, assuming the existence of such streams of particles, and disregarding such actions as might be due to the properties, chemical or physical, of the projected matter, we have to consider the following specific actions:

First.. There is the thermal effect. The temperature of the electrode or impact body does not in any way give us an idea of the degree of heat of the particles, but, if we consider the probable velocities only, they correspond to temperatures which may be as high as 100,000 degrees centigrade. It may be sufficient that the particles are simply at a high temperature to produce an injurious action, and in fact, many evidences point in this direction. But against this is the experimental fact that we cannot demonstrate such a transference of heat, and no satisfactory explanation is found yet, although, in carrying my investigations in this direction, I have arrived at some results.

Second, there is the purely electrical effect. We have absolute experimental evidence that that particles or rays, to express myself generally, convey an immense amount of electricity, and I have even found a way of how to estimate and measure that amount. Now it is likewise possible that the mere fact of these particles being highly electrified is sufficient to cause the destruction of the tissue. Certainly, on contact with the skin, the electrical charges will be given off, and may give rise to strong and destructive local currents in minute paths of the tissue. Experimental results are in accord with this view, and, in pushing my inquiry in this direction, I have been still more successful than in the first. Yet, while as I have suggested before, this view explains best the action on a sensitive layer, experiment shows that, when the supposed particles traverse a grounded plate, they are not deprived entirely of their electrification, which is not satisfactorily explained.

The *third* effect to be considered is the electro-chemical. The charged particles give rise to an abundant generation of ozone and other gases, and these we know, by experiment, destroy even such a thing as rubber, and are, therefore, the most likely agent in the destruction of the skin, and the evidences are strongest in this direction, since a small layer of a fluid, preventing the contact of gaseous matter with the skin, seems to stop all action.

The *last* effect to be considered is the purely mechanical. It is thinkable that material particles, moving with great speed, may, merely by a mechanical impact and unavoidable heating at such speeds, be sufficient to deteriorate the tissues, and in such a case deeper layers might also be injured, whereas it is very probable that no such thing would occur if any of the former explanations would be found to hold.

Summing up my experimental experiences and the conclusions derived from them, it would seem advisable, first, to abandon the use of bulbs containing platinum; second, to substitute for them a properly constructed Lenard tube, containing pure aluminum only, a tube of this kind having, besides, the advantage that it can be constructed with great mechanical precision, and therefore is capable of producing much sharper impressions; third, to use a protecting screen of aluminum sheet, as suggested, or, instead of this, a wet cloth or layer of a fluid; fourth, to make the exposures at distances of, at least, 14 inches, and preferably expose longer at a larger distance.

New York, May 1, 1897.

Section III

THE SOURCE OF ROENTGEN RAYS AND THE PRACTICAL CONSTRUCTION AND SAFE OPERATION OF LENARD TUBES.

To the Editor of Electrical Review:

I have for some time felt that a few indications in regard to the practical construction of Lenard tubes of improved designs, a great number of which I have recently exhibited before the New York Academy of Sciences [April 6, 1897], would be useful and timely, particularly as by their proper construction and use much of the danger attending the experimentation with the rays may be avoided. The simple precautions which I have suggested in my previous communications to your esteemed journal are seemingly disregarded, and cases of injury to patients are being almost daily reported, and in view of this only, were it for no other reason, the following lines, referring to this subject, would have been written before had not again pressing and unavoidable duties prevented me from doing so. A short and, I may say, most unwelcome interruption of the work which has been claiming my attention makes this now possible. However, as these opportunities are scarce, I will utilize the present to dwell in a few words on some other matters in connection with this subject, and particularly on a result of importance which I have reached some time ago by the aid of such a Lenard tube, and which, if I am correctly informed, I can only in part consider as my own, since it seems that practically it has been expressed in other words by Professor Roentgen in a recent communication to the Academy of Sciences of Berlin. The result alluded to has reference to the much disputed question of the source of the Roentgen rays. As will be remembered, in the first announcement of his discovery, Roentgen was of the opinion that the rays which affected the sensitive layer emanated from the fluorescent spot on the glass wall of the bulb; other scientific men next made the cathode responsible; still others the anode, while some thought that the rays were emitted solely from fluorescent powders or surfaces, and speculations, mostly unfounded, increased to such an extent that, despairingly, one would exclaim with the poet:

> "O glücklich, wer noch hoffen kann,
> Aus diesem Meer des Irrtums aufzutauchen!"[*]

My own experiments led me to recognize that, regardless of the location, the chief source of these rays was the place of the *first* impact of the projected stream of particles within the bulb. This was merely a broad statement, of which that of Professor Roentgen was a special case, as in his first experiments the fluorescent spot on the glass wall was, incidentally, the place of the first impact of the cathodic stream. Investigations carried on up to the present day have only confirmed the correctness of the above opinion, and the place of the first collision of the stream of particles—be it an anode or independent impact body, the glass wall or an aluminum window—is still found to be the principal source of the rays. But as will be seen presently, it is not the only source.

Since recording the above fact my efforts were directed to finding answers to the following questions: First, is it necessary that the impact body should be within the tube? Second, is it required that the obstacle in the path of the cathodic stream should be a solid or liquid? And, third, to what extent is the velocity of the stream necessary for the generation of and influence upon the character of the rays emitted?

In order to ascertain whether a body located outside of the tube and in the path or in the direction of the stream of particles was capable of producing the same peculiar phenomena as an object located inside, it appeared necessary to first show that there is an actual penetration of the particles through the wall, or otherwise that the actions of the supposed streams, of whatever nature they might be, were sufficiently pronounced in the outer region close to the wall of the bulb as to produce some of the effects which are peculiar

* Oh, happy he who still hopes to rise,
Out of this sea of errors and false views! *Faust*, 1064-5

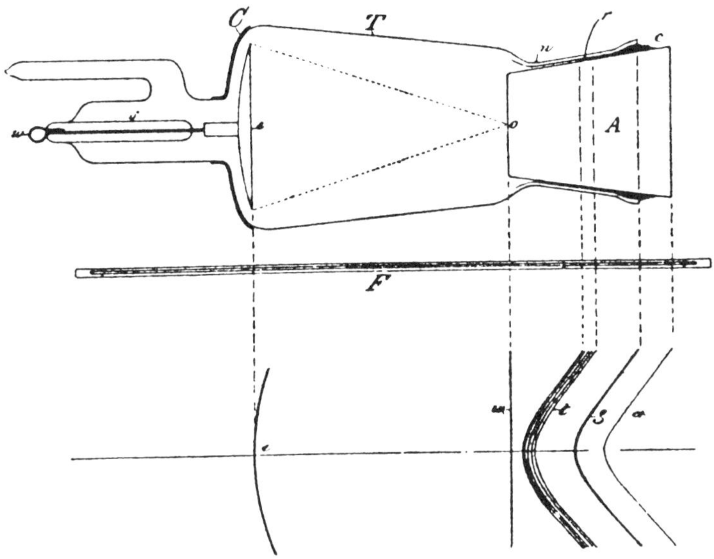

Fig. 1. Illustrating an experiment revealing the real source of the Roentgen rays.

to a cathodic stream. It was not difficult to obtain with a properly prepared Lenard tube, having an exceedingly thin window, many and at first surprising evidences of this character. Some of these have already been pointed out, and it is thought sufficient to cite here one more which I have since observed. In the hollow aluminum cap *A* of a tube as shown in diagram Fig. 1, which will be described in detail, I placed a half-dollar silver piece, supporting it at a small distance from and parallel to the window or bottom of the cap by strips of mica in such a manner that it was not touching the metal of the tube, an air space being left all around it. Upon exciting the bulb for about 30 to 45 seconds by the secondary discharge of a powerful coil of a novel type now well known, it was found that the silver piece was rendered so hot as to actually scorch the hand; yet the aluminum window, which offered a very insignificant obstacle to the cathodic stream, was only moderately warmed. Thus it was shown that the silver alloy, owing to its density and thickness, took up most of the energy of the impact, being acted upon by the particles almost identically as if it had been inside of the bulb, and what is more, indications were obtained, by observing the shadow, that it behaved like a second source of the rays, inasmuch as the outlines of the shadows, instead of being sharp and clear as when the half-dollar peace was removed, were dimmed. It was immaterial for the chief object of the inquiry to decide by more exact methods whether the cathodic particles actually penetrated the window, or whether a new and separate stream was projected from the outer side of the window. In my mind there exists not the least doubt that the former was the case, as in this respect I have been able to obtain numerous additional proofs, upon which I may dwell in the near future.

I next endeavored to ascertain whether it was necessary that the obstacle outside was, as in this case, a solid body, or a liquid, or broadly, a body of measurable dimensions, and it was in investigating in this direction that I came upon the important result to which I referred in the introductory statements of this communication. I namely observed rather accidentally, although I was following up a systematic inquiry, what is illustrated in diagram Fig. 1. The diagram shows a Lenard tube of improved design, consisting of a tube *T* of thick glass tapering towards the open end, or neck *n*, into which is fitted an aluminum cap *A*, and a spherical cathode *e*, supported on a glass stem *s*, and platinum wire *w* sealed in the opposite end of the tube as usual. The aluminum cap *A*, as will be observed, is not in actual contact with the ground glass wall, being held at a small distance from the latter by a narrow and continuous ring of tinfoil *r*. The outer space between the glass and the cap *A* is filled with cement *c*, in a manner which I shall later describe. *F* is a Roentgen screen such as is ordinarily used in making the observations.

Now, in looking upon the screen in the direction from *F* to *T*, the dark lines indicated on the lower part of the diagram were seen on the illuminated background. The curved line *e* and the straight line *W* were, of course, at once recognized as the outlines of the cathode *e* and the bottom of the cap *A* respectively, although, in consequence of a confusing optical illusion, they appeared much closer together than they actually were. For instance, if the distance between *e* and *o* was 5 inches, these lines would appear on the screen about 2 inches apart, as nearly as I could judge by the eye. This illusion may be easily explained and is quite unimportant, except that it might be of some moment to physicians to keep this fact in mind when making examinations with the screen as, owing to the above effect, which is sometimes exaggerated to a degree hard to believe, a completely erroneous idea of the distance of the various parts of the object under examination might be gained, to the detriment of the surgical operation. But while the lines *e* and *W* were easily accounted for, the curved lines *t, g, a* were at first puzzling. Soon, however, it was ascertained that the faint line *a* was the shadow of the edge of the aluminum cap, the much darker line *g* that of the rim of the glass tube *T*, and *t* the shadow of the tinfoil ring *r*. These shadows on the screen *F* clearly showed that the agency which affected the fluorescent material was proceeding from the space outside of the bulb towards the aluminum cap, and chiefly from the region through which the primary disturbances or streams emitted from the tube through the window were passing, which observation could not be explained in a more plausible manner than by assuming that the air and dust particles outside, in the path of the projected streams, afforded an obstacle to their passage and gave rise to impacts and collisions spreading through the air in all directions, thus producing continuously new sources of the rays. It is this fact which, in his recent communication before mentioned, Roentgen has brought out. So, at least, I have interpreted his reported statement that the rays emanate from the irradiated air. It now remains to be shown whether the air, from which carefully all foreign particles are removed, is capable of behaving as an impact body and source of the rays, in order to decide whether the generation of the latter is dependent on the presence in the air of impact particles of measurable dimensions. I have reasons to think so.

With the knowledge of this fact we are now able to form a more general idea of the process of generation of the radiations which have been discovered by Lenard and Roentgen. It may be comprised in the statement that the streams of minute material particles projected from an electrode with great velocity in encountering obstacles wherever they may be, within the bulb, in the air or other medium or in the sensitive layers themselves, give rise to rays or radiations possessing many of the properties of those known as light. If this physical process of generation of these rays is undoubtedly demonstrated as true, it will have most important consequences, as it will induce physicists to again critically examine many phenomena which are presently attributed to transverse ether waves, which may lead to a radical modification of existing views and theories in regard to these phenomena, if not as to their essence so, at least, as to the mode of their production.

My effort to arrive at an answer to the third of the above questions led me to the establishment, by actual photographs, of the close relationship which exists between the Lenard and Roentgen rays. The photographs bearing on this point were exhibited at a meeting of the New York Academy of Sciences—before referred to—April 6, 1897, but unfortunately, owing to the shortness of my address, and concentration of thought on other matters, I omitted what was most important; namely, to describe the manner in which these photographs were obtained, an oversight which I was able to only partially repair the day following. I did, however, on that occasion illustrate and describe experiments in which was shown the deflectibility of the Roentgen rays by a magnet, which establishes a still closer relationship, if not identity, of the rays named after these two discoverers. But the description of these experiments in detail, as well as of other investigations and results in harmony with and restricted to the subject I brought before that scientific body, will appear in a longer communication which I am slowly preparing.[*]

* The "longer communication" referred to is undoubtedly the lecture, "High Frequency Oscillation for Electro-Therapeutics and Other Purposes," presented before The American Electro-Therapeutics Association, Sept. 14, 1898.

To bring out clearly the significance of the photographs in question, I would recall that, in some of my previous contributions to scientific societies, I have endeavored to dispel a popular opinion before existing that the phenomena known as those of Crookes were dependent on and indicative of high vacua. With this object in view, I showed that phosphorescence and most of the phenomena in Crookes bulbs were producible at greater pressures of the gases in the bulbs by the use of much higher or more sudden electromotive impulses. Having this well demonstrated fact before me, I prepared a tube in the manner described by Lenard in his first classical communication on this subject. The tube was exhausted to a moderate degree, either by chance or of necessity, and it was found that, when operated by an ordinary high-tension coil of a low rate of change in the current, no rays of any of the two kinds could be detected, even when the tube was so highly strained as to become very hot in a few moments. Now, I expected that, if the suddenness of the impulses through the bulb were sufficiently increased, rays would be emitted. To test this I employed a coil of a type which I have repeatedly described, in which the primary is operated by the discharges of a condenser. With such an instrument any desired suddenness of the impulses may be secured, there being practically no limit in this respect, as the energy accumulated in the condenser is the most violently explosive agent we know, and any potential or electrical pressure is obtainable. Indeed, I found that in increasing the suddenness of the electromotive impulses through the tube—without, however, increasing, but rather diminishing the total energy conveyed to it—phosphorescence was observed and rays began to appear, first the feebler Lenard rays and later, by pushing the suddenness far enough, Roentgen rays of great intensity, which enable me to obtain photographs showing the finest texture of the bones. Still, the same tube, when again operated with the ordinary coil of a low rate of change in the primary current, emitted practically no rays, even when, as before stated, much more energy, as judged from the heating, was passed through it. This experience, together with the fact that I have succeeded in producing by the use of immense electrical pressures, obtainable with certain apparatus designed for this express purpose, some impressions in free air, have led me to the conclusion that in lightning discharges Lenard and Roentgen rays must be generated at ordinary atmospheric pressure.

At this juncture I realize, by a perusal of the preceding lines, that my scientific interest has dominated the practical, and that the following remarks must be devoted to the primary object of this communication—that is, to giving some data for the construction to those engaged in the manufacture of the tubes and, perhaps, a few useful hints to practicing physicians who are dependent on such information. The foregoing was, nevertheless, not lost for this object, inasmuch as it has shown how much the result obtained depends on the proper construction of the instruments, for with ordinary implements, most of the above observations could not have been made.

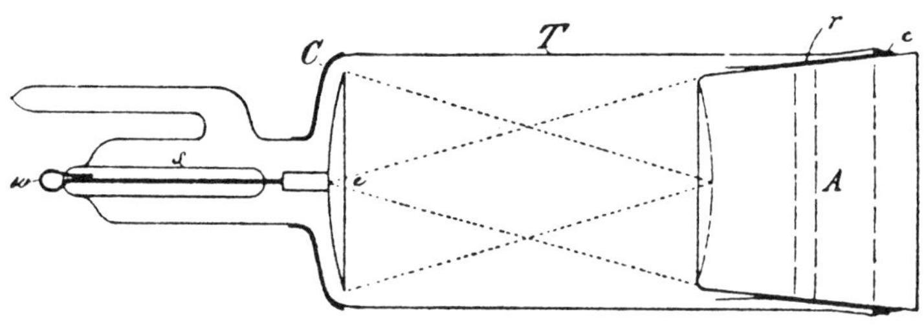

Fig. 2.—Improved Lenard tube.

I have already described the form of tube illustrated in Fig. 1, and in Fig. 2 another still further improved design is shown. In this case the aluminum cap A, instead of having a straight bottom as before, is shaped spherically, the center of the sphere coinciding with that of the electrode *e*, which itself, as in Fig. 1, has its focus in the center of the window of cap *A*, as indicated by the dotted lines. The aluminum cap *A* has a tinfoil ring *r*, as that in Fig. 1, or else the metal of the cap is spun out on that place so as to afford a bearing of small surface between the metal and the glass. This is an important practical detail as, by making the bearing surface small, the pressure per unit of area is increased and a more perfect joint made. The ring *r* should be first spun out and then ground to fit the neck of the bulb. If a tinfoil ring is

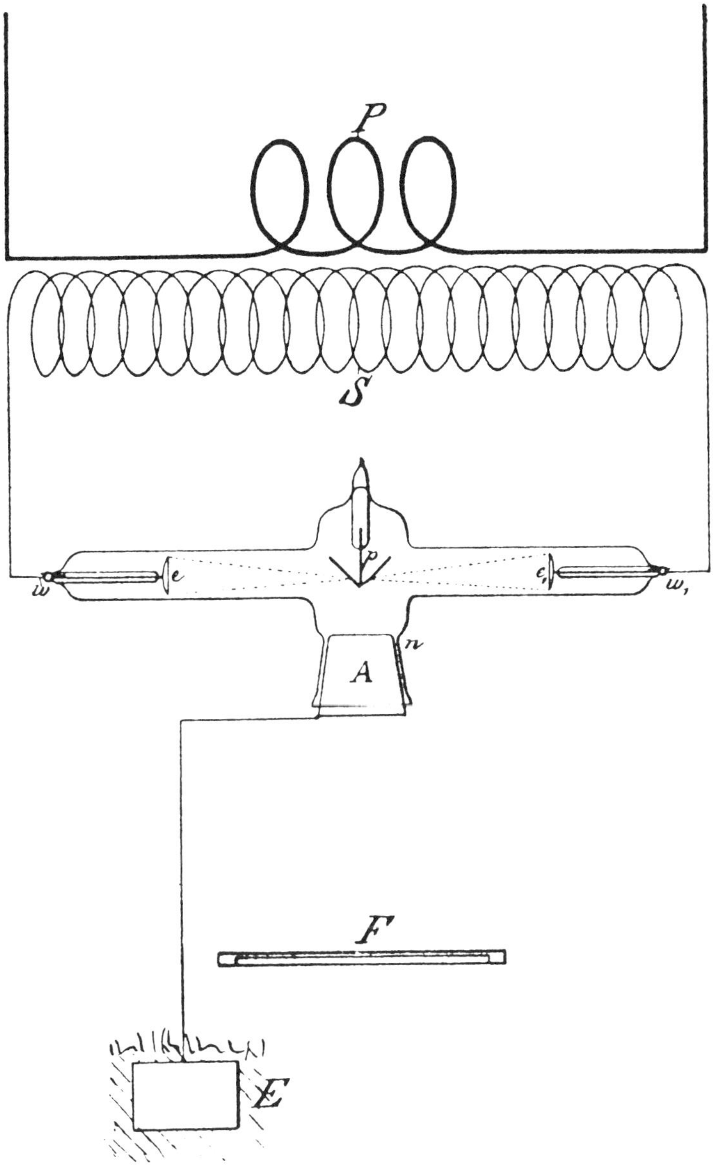

Fig. 3.—Illustrating arrangement with improved double-focus tube for reducing injurious actions.

used instead, it may be cut out of one of the ordinary tinfoil caps obtainable in the market, care being taken that the ring is very smooth.

In Fig. 3 I have shown a modified design of tube which, as the two types before described, was comprised in the collection I exhibited. This, as will be observed, is a double-focus tube, with impact plates of iridium alloy and an aluminum cap *A* opposite the same. The tube is not shown because of any originality in design, but simply to illustrate a practical feature. It will be noted that the aluminum caps in the tubes described are fitted inside of the necks and not outside, as is frequently done. Long experience as demonstrated that it is practically impossible to maintain a high vacuum in a tube with an outside cap. The only way I have been able to do this in a fair measure is by cooling the cap by a jet of air, for instance, and observing the following precautions: The air jet is first turned on slightly and upon this the tube is excited. The current through the latter, and also the air pressure, are then gradually increased and brought to the normal working condition. Upon completing the experiment the air pressure and current through the tube are both gradually reduced and both so manipulated that no great differences in temperature result between the glass and aluminum cap. If these precautions are not observed the vacuum will be immediately impaired in consequence of the uneven expansion of the glass and metal.

With tubes as these presently described, it is quite unnecessary to observe this precaution if proper care is taken in their preparation. In inserting the cap the latter is cooled down as low as it is deemed advisable without endangering the glass, and it is then gently pushed in the neck of the tube, taking care that it sets straight.

The two most important operations in the manufacture of such a tube are, however, the thinning down of the aluminum window and the sealing in of the cap. The metal of the latter may be one thirty-second or even one-sixteenth of an inch thick, and in such case the central portion may be thinned down by a countersink tool about one-fourth of an inch in diameter as far as it is possible without tearing the sheet. The further thinning down may then be done by hand with a scraping tool; and, finally, the metal should be gently beaten down so as to surely close the pores which might permit a slow leak. Instead of proceeding in this way I have employed a cap with a hole in the center, which I have closed with a sheet of pure aluminum a few thousandths of an inch thick, riveted to the cap by means of a washer of thick metal, but the results were not quite as satisfactory.

In sealing the cap I have adopted the following procedure: The tube is fastened on the pump in the proper position and exhausted until a permanent condition is reached. The degree of exhaustion is a measure of perfection of the joint. The leak is usually considerable, but this is not so serious a defect as might be thought. Heat is now gradually applied to the tube by means of a gas stove until a temperature up to about the boiling point of sealing wax is reached. The space between the cap and the glass is then filled with sealing wax of good quality; and, when the latter begins to boil, the temperature is reduced to allow its settling in the cavity. The heat is then again increased, and this process of heating and cooling is repeated several times until the entire cavity, upon rcduction of the temperature, is found to be filled uniformly with the wax, all bubbles having disappeared. A little more wax is then put on the top and the exhaustion carried on for an hour or so, according to the capacity of the pump, by application of moderate heat much below the melting point of the wax.

A tube prepared in this manncr will maintain the vacuum very well, and will last indefinitely. If not used for a few months, it may gradually lose the high vacuum, but it can be quickly worked up. However, if after long use it becomes necessary to clean the tube, this is easily done by gently warming it and taking off the cap. The cleaning may be done first with acid, then with highly diluted alkali, next with distilled water, and finally with pure rectified alcohol.

These tubes, when properly prepared, give impressions much sharper and reveal much more detail than those of ordinary make. It is important for the clearness of the impressions that the electrode should be properly shaped, and that the focus should be exactly in the center of the cap or slightly inside. In fitting in the cap, the distance from the electrode should be measured as exactly as possible. It should also be remarked that the thinner the window, the sharper are the impressions, but it is not advisable to make it too thin, as it is apt to melt at a point on turning on the current.

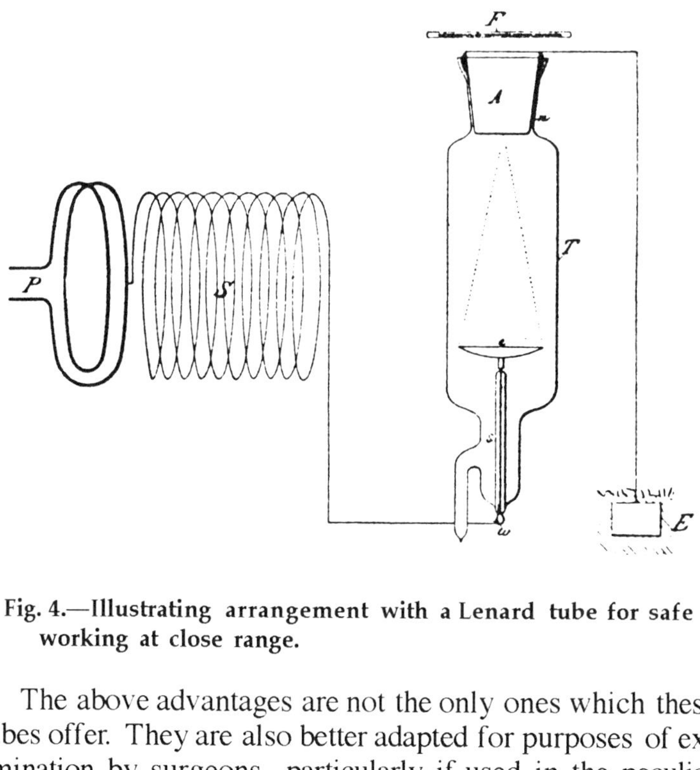

Fig. 4.—Illustrating arrangement with a Lenard tube for safe working at close range.

The above advantages are not the only ones which these tubes offer. They are also better adapted for purposes of examination by surgeons, particularly if used in the peculiar manner illustrated in diagrams Fig. 3 and Fig. 4, which are self-explanatory. It will be seen that in each of these the cap is connected to the ground. This decidedly diminishes the injurious action and enables also to take impressions with very short exposures of a few seconds only at close range, inasmuch as, during the operation of the bulb, one can easily touch the cap without any inconvenience, owing to the ground connection. The arrangement shown in Fig. 4 is particularly advantageous with a form of single terminal, which coil I have described on other occasions and which is diagrammatically illustrated, P being the primary and S the secondary. In this instance the high-potential terminal is connected to the electrode, while the cap is grounded. The tube may be placed in the position indicated in the drawing, under the operating table and quite close, or even in contact with the body of the patient, if the impression requires only a few seconds as, for instance, in examining parts of the members. I have taken many impressions with such tubes and have observed no injurious action, but I would advise not to expose for longer than 2 or 3 minutes at very short distances. In this respect the experimenter should bear in mind what I have stated in previous communications. At all events it is certain that, in proceeding in the manner described, additional safety is obtained and the process of taking impressions much quickened. To cool the cap, a jet of air may be used, as before stated, or else a small quantity of water may be poured in the cap each time when an impression is taken. The water only slightly impairs the action of the tube, while it maintains the window at a safe temperature. I may add that the tubes are improved by providing back of the electrode a metallic coating C, shown in Fig. 3 and Fig 4.[*]

New York, August 9, 1897.

* These figure references are incorrect in the original text of this communication; they should be to Fig. 1 and Fig. 2.

Appendix

Contemporary Reviews of the Lecture by Nikola Tesla Before the New York Academy of Sciences April 6, 1897

Electrical Engineer, April 14, 1897

"Mr. Tesla on Roentgen Rays" (Editorial) p. 398, and "Mr. Tesla Before the New York Academy of Sciences" (Review) pp. 400-401.

Electrical Review (N.Y.), April 14, 1897

"The New York Academy of Science. An Interesting Electrical Exhibition—Address by Nikola Tesla announcing recent achievements" (Editorial) p. 175.

THE

ELECTRICAL ENGINEER

[INCORPORATED.]

PUBLISHED EVERY WEDNESDAY

120 Liberty Street, New York.

Telephone: 1323 Cortlandt. Cable Address: LENGINEER.

T. COMMERFORD MARTIN AND JOSEPH WETZLER, Editors.

A. C. SHAW, Secy. and Business Manager.

WESTERN OFFICE - - - 1564 Monadnock Block, Chicago, Ill
PHILADELPHIA OFFICE - - - - - 916 Betz Building.

Terms of Subscription

United States, Canada and Mexico	per year.	$3.00
Four or more Copies in Clubs (each)	"	2.50
Great Britain and other Foreign Countries within the Postal Union	"	5.00
Single Copies		.10

[*Entered as second-class matter at the New York Post Office, April 9, 1888.*]

VOL. XXIII. NEW YORK, APRIL 14, 1897. No. 467.

MR. TESLA ON ROENTGEN RAYS.

LAST week Mr. Tesla gave some of the results of his experiments in the domain of Röntgen rays, before the New York Academy of Sciences, and showed some of his latest types of high frequency generators. Mr. Tesla still adheres to his original view that the Röntgen effects are due to the action of molecules projected from the tube at high velocities, but we must confess our inability to reconcile this view with the results of his experiments in deflecting the Röntgen rays by means of a magnet, unless we assume the molecules charged and at the same time endowed with a vortical motion, a point which Mr. Tesla did not elaborate on. It is so rarely that Mr. Tesla appears on the lecture platform that it is to be regretted that the conditions surrounding his last appearance were not more favorable. It was hardly fair to Mr. Tesla, or to the large audience which had assembled solely to greet him, to delay his appearance until two other estimable speakers had taken up the time of the audience for nearly three-quarters of an hour. The result was that it was close on to 10 o'clock before Mr. Tesla began, and he was forced to conclude his address in its initial stages. It is to be hoped, for the benefit of the science at large that Mr. Tesla will find time to write out his address in full for the Transactions of the Academy. Brief as his utterances were, they were extremely interesting and they evidenced a mellowing and a mastery indicative of higher perfection than ever of his powers as an investigator and elucidator of obscure natural phenomena.

MR. TESLA BEFORE THE NEW YORK ACADEMY OF SCIENCES.

ON April 6, Mr. Nikola Tesla delivered before the New York Academy of Sciences an address "On the Streams of Lenard and Röntgen, with Novel Apparatus for Their Production." The lecturer began by stating that in 1894, in experimenting to determine the actinic action of phosphorescent light emanating from vacuum tubes, he had found that the actinic power of Crookes tubes varied greatly, and in the most anomalous manner. Thus some tubes emitting a strong light had very little action on photographic plates, while some showing a feeble light acted very strongly on such plates. A large number of these plates, made by Tonnele for the Century Magazine, were stored in Mr. Tesla's laboratory when it was destroyed by fire, and thus he was unable now to examine them for the Röntgen effects. He believed that he had just missed the discovery which made Röntgen famous and though he was thoroughly familiar with Lenard's work, he did not see far enough.

Mr. Tesla then reviewed his work in the direction of obtaining a reliable apparatus for generating high frequency currents, which he recognized as the keynote to the production of vacuum tube lighting. He told how he had met with difficulties at every point; how a small bubble of air would destroy the value of the coil, or how one-quarter of an inch of wire too much or too little would throw a coil out of balance; how one day a coil would run cold and on another day hot, etc. By finally calling to his aid the condenser, Mr. Tesla stated, he had succeeded in obtaining the desired action and now nothing stood in the way of securing millions of vibrations from ordinary circuits. The lecturer then briefly explained the principle of the condenser discharge as applied to high frequency currents, and pointed to various types of apparatus on the lecture table designed to utilize the principle.

In order to demonstrate the action of the high frequency currents Mr. Tesla attached a circular loop of heavy copper wire to the terminals of a high frequency generator and

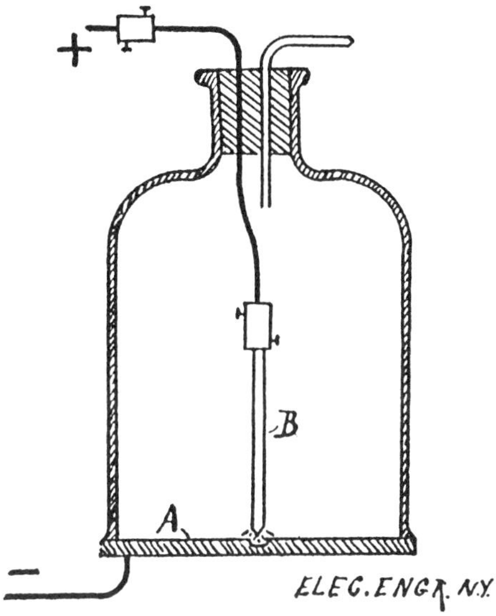

TESLA'S ARC X-RAY GENERATOR.

brought to brilliant incandescence a small lamp connected diametrically across the heavy loop; the illumination of the lamp could be varied by connecting it to various points on the diameter of the loop. Mr. Tesla also lit the lamp by the current induced in a second loop brought in proximity to the first. He explained that the coil accomplishing this work had a resistance of 600 ohms and an inductance of 6 henrys; the coil was connected to the circuit one-half the time and took from the primary circuit only 5 watts.

Mr. Tesla next showed a vacuum tube lit from the terminals of another high frequency machine. He also showed a coil consisting of a single turn of heavy wire which formed the core, as it were, of a small coil of a few turns wound on a paper cylinder surrounding the heavy wire. When connected to a high frequency generator, this apparatus gave a 4-inch spark with an expenditure of energy equal to that taken by one lamp. Mr. Tesla stated that his high frequency generators were so constructed that they could be connected to any existing circuits, direct or alternating.

Returning to the Röntgen rays, the lecturer stated that he had succeeded in discovering a new source of these rays, far more powerful than any heretofore available, though the difficulty of maintaining it was very great. This new source is the electric arc; not the ordinary arc, however. The arc required for the purpose is that maintained between a platinum terminal and an aluminum plate, as illustrated in the accompanying diagram, where A represents the aluminum plate and B the platinum, enclosed in a glass jar.

Mr. Tesla stated that he had also succeeded in deflecting the Röntgen rays by a magnet. He had proved this by deflecting the rays into a condenser placed a long distance from the source of the rays, and which in 5 seconds was charged sufficiently to throw a galvanometer needle off the scale.

Electrical Review

Times Building, 41 Park Row,

P. O. Box 2339. NEW YORK.

EDITOR:
CHARLES W. PRICE.

ASSOCIATE EDITOR:
STEPHEN L. COLES.

REGISTERED CABLE ADDRESS:
"ELECTVIEW," NEW YORK.

1646 "CORTLANDT," TELEPHONE CALL.

WESTERN OFFICE:
CARL E. KAMMEYER, Representative,
1627 MONADNOCK BLOCK, CHICAGO.

LONDON OFFICE:
HENRY W. HALL, Representative,
42 OLD BROAD STREET, LONDON, E. C.

News, Subscriptions and Advertisements.

Subscription to the

ELECTRICAL REVIEW.

One Year, United States and Canada, -	$3.00
One Year, Foreign Countries, - -	5.00
Single Copy, - - - - - - -	.10

Copies can only be supplied for one year back at 25 cents each.

Special Notice to Advertisers.

CHANGES for advertisements must be in this office by Friday noon for the following week's issue.

NEW ADVERTISEMENTS should be in the office not later than Saturday noon to insure publication the following week.

NEW YORK, MAY 5, 1897.

THE NEW YORK ACADEMY OF SCIENCE.

AN INTERESTING ELECTRICAL EXHIBITION — ADDRESS BY NIKOLA TESLA, ANNOUNCING RECENT ACHIEVEMENTS.

The New York Academy of Sciences gave its fourth annual reception and exhibit of recent progress in science in the American Museum of Natural History, New York, April 5 and 6.

In the division devoted to chemistry, in charge of Charles A. Doremus, were exhibited a number of electric furnaces, and in the electrical division, in charge of G. F. Sever, were to be seen a number of historical collections of incandescent and arc lamps, instruments of precision, Crookes tubes, induction coils, etc. Mr. W. H. Meadowcroft exhibited a Thomson 14 inch spark inductorium, and tubes and fluoroscopes, and Dr. M. I. Pupin had on view an induction coil and circuit breaker for the generation of Roentgen rays. Dr. Max Osterberg exhibited some specimens of German tubes, and Dr. Doremus showed one of the original Henry induction coils.

Tuesday evening a large audience assembled to listen to an address by Nikola Tesla on "The Streams of Lenard and Roentgen and Novel Apparatus for their Production". The lecturer had on exhibition a number of his perfected electrical oscillators, designed particularly to be used in place of the induction coils of old type. He exhibited numerous diagrams illustrating results obtained by him in his experiments, to ascertain the nature of the Roentgen rays. There were over 100 different working drawings of bulbs, showing the extent of the ground covered by the lecturer in this field. There were, besides, a number of remarkable photographs exhibited, which were taken by means of one of the small oscillators exhibited. These instruments and photographs were examined with the keenest interest after the lecture by a large number who remained and insisted on a supplementary lecture, which Mr. Tesla kindly consented to give, notwithstanding that, as he playfully remarked, he had not yet had his dinner.

Mr. Tesla introduced the subject of his lecture by stating that at the close of 1894 he investigated the actinic power of phosphorescent bodies, and during these experiments, which were carried on by the assistance of Tonnelé & Company, photographers of this city, working for the *Century Magazine*, a great number of plates showed curious marks and defects, which were noted both by the speaker and the photographers, but not recognized as being due to the Roentgen rays. Just as he was beginning to look into the nature of the phenomena his laboratory was burned, and before his work was thoroughly resumed in his new laboratory Roentgen's discovery was announced.

Mr. Tesla first described the instruments exhibited, dwelling on the principles underlying them, and he illustrated, in a few striking experiments, the practical and economical features of the machines. The greatest interest seemed to be evinced by the audience when he lighted brilliantly a vacuum tube with one of his machines, giving, as he said, 800,000 vibrations a second. What rendered them particularly valuable, he stated, was that they worked equally well on alternating or direct-current circuits, and transform these currents with high economy into high frequency electrical vibrations.

Mr. Tesla announced two important discoveries relating to the Roentgen rays. First, he said he had discovered a new and powerful source of the rays in an electric arc, formed under peculiar conditions. The second discovery was the deflection of the Roentgen rays by means of a magnet. This discovery is particularly important in establishing the identity of the Roentgen rays, and those discovered by Lenard in 1891, and is, therefore, one of the most valuable contributions to our knowledge of these rays.

Tesla dwelt also on some phenomena of rotation of bodies in bulbs, and, among other things, described his method of determining exactly the speed of a dynamo when taking readings.

To a few interested scientific men Mr. Tesla showed a great number of diagrams illustrating experiments he had performed which tended to prove the correctness of the views he holds in regard to the Roentgen phenomena being caused by material particles projected with great velocity.

The ELECTRICAL REVIEW hopes to be able to present in full, in an early issue, Mr. Tesla's lecture.

Acknowledgments

I am indebted to James Hardesty, historian specializing in early static machines and X-ray apparatus, for the inspiration to assemble this record and for reviewing the final manuscript. I wish to thank James Corum and Kenneth Corum for their invaluable exchanges during the years preceding the manuscript's completion as well as its final review. In addition, I also wish to thank reference librarians Sue Christensen, Bobbi Jones, and David Schmit together with chief Lynn Taylor and her always tolerant Interlibrary Loan Department staff, Denver Public Library system, for their generous assistance; Roger Boyle for typographic production; Dave Allinger for photographic imaging and processing; and last but surely not least, Gary Peterson of Twenty First Century Books for the continued encouragement he has given.

L.I.A.

Sponsorship

Dynamic Systems, Inc., Leicester, North Carolina
Future Free Transportation, Hamer, Idaho

Index

Abrahams, Harold, and Savin, Marion, 4n, 27

Absorption loss, 45

Actinic actions, 31, 112, 116

Actions; see Effects

Activity series of metals, 22

Alley, Dickson D., 31

Allied powers, 16

Alternating current, municipal power, 4, 9, 15, 27, 113

Aluminum, use of, xvii, 20, 23, 65, 84-87, 90, 92, 94, 96-99, 102, 104, 105

American Electro-Therapeutic Association, 5, 6, 100n

American Institute of Electrical Engineers (AIEE), *x*, 4, 6

American Museum of Natural History, 115

Ampère, André Marie, *xiv*

AND logic gate, 9

Anderson, Leland I., *xvi*, *xvii*, 1n

Anode, 16n, 18, 21, 23, 95, 96
  virtual, 25

Baltimore Lectures on Molecular Dynamics and the Wave Theory of Light, 21n

Berlin Academy of Sciences, 95

Black-body radiation, *xiii*

Boscović, Rudjer, 20, 20n, 21
  "force curve," 21

Bragg, Sir William. H., 22

Break; see Tesla coil, primary

Bremsstrahlung, 23

British Association for the Advancement of Science, 9, 10

Cathode, 16n, 18, 21, 23-25, 86, 95, 98, 99
  dark space; see Dark space
  rays, *xiv*, *xv*
  stream(s), *xvii*, *xvii*, 86, 87, 92, 96-98

Cathodic matter, 92

Century Magazine, 6, 8n, 10, 31, 112, 116

Coles, Stephen L., 6

Columbian Exposition, *xi*

Compton, Arthur H., *xviii*, 22, 22n

Condenser discharges, 9, 14, 34, 35, 43, 45, 47, 48, 51, 54, 55, 68, 101, 112

Controller, circuit ; see Tesla coil, primary

Cooper Union, 7

Cooper, Peter, 7

Cornell Dubilier Co., 12

Corum, James F. and Kenneth L., *xvii*, 24n

Crookes, Sir William, *x*, *xiv*, 18
  bulb(s)/tube(s), xvi, 20, 31, 83, 101, 112, 115
  dark space; see Dark space
  phenomena, 101

Dana, Charles A., 7

Dark space:
  cathode, 19
  Crookes, 19
  Faraday, 19
  Hittorff, 19

Diode(s):
  open-air, *xvii*
  relativistic electron-beam (REB), *xvii*

Direct current, municipal power, 4, 9, 43, 45, 68, 113

Discharge circuit; see Tesla coil, primary

Dollard, Eric P., 25

Doremus, Charles A., 115

Eberhart, N.M., 9n

Edgerton, Harold E., 17

Effects of material stream emanations (see also Injury), 93, 94

Electrical Engineer, 1n, 3, 3n, 4-6, 22n

Electrical Engineering, 3, 6

Electrical Experimenter, 75n

Electrical Review (London), *xvi*, 1n

Electrical Review (N.Y.), *xvi*, *xvii*, 1n, 2, 3, 5, 7, 7n, 10, 15, 16, 19, 21, 22, 23, 23n, 24n, 25, 26, 32, 83, 95

Electrical World, 3, 4, 6, 27n

Electrician, 3

Electricity, 3

Electron beams/rays (see also Lenard rays), *xiv*, 20, 23-25, 101

Electrostatic:
  action/influence, 84, 85
  screen, 84, 85, 87, 90, 92

Electrotherapeutics, 5, 6n, 9

Energy conversion efficiency, 9, 33, 43

Engineering Magazine, 3, 4

Ether, *xiv*
  waves, *xiii*, 100

Exposure (Lenard and Roentgen rays),
  distance, 5n, 23, 85, 86, 88-92, 94, 106, 107
  duration, 10, 89-92, 106, 107

Faraday, Michael, *xiv*
  dark space; see Dark space

Fluorescent:
  material, *xiii*, *xv*, 99
  spot, bulb, 96

Fluorescent-emulsion films, 22

Fluorescent/phosphorescent lamps, 3, 10, 11, 31, 44, 75, 112, 115, 116

Franklin Institute, *x*

Geissler, Heinrich:
  tube, 8
  vacuum pump, *xiv*

General Electric Co., 26

Glass, insulation vs. temperature, 25

Goldstein, Eugen, *xiv*

Greene, William H., Dr., 27

Guericke, Otto von, *xiii*

Hawksbee, Francis, *xiv*

Henry, Joseph, *x*, *xiv*

Hertz, Heinrich, `

Hewitt, Edward R., 7, 7n, 8

High Frequency Electric Currents in Medicine and Dentistry , 9n

High frequency oscillators/resonators, 2, 5, 6n, 11, 16, 34, 40, 53, 71, 74n, 112, 113, 116

High voltage:
  alternating current(s), 8, 13, 15, 24, 24n
  apparatus, 16
  oscillators, 11

High-field emission, 24

Hittorff, Jacques, *xiv*
  dark space; see Dark space

Hopkinson, Mr., 10

Hurtful actions (Lenard and Roentgen rays); see X-ray apparatus, harms

Imaging (photographs, shadow graphs), 5n, 7, 8, 10, 23, 29, 91, 99, 101, 105-107, 118

Impact:
  body/plate, 20, 21, 23, 85, 86, 93, 96, 99, 100, 102, 104
  particles, 93, 94, 96, 98-100

Individualization, art of, 9

Injury (Lenard and Roentgen rays); see X-ray apparatus, harms

Institution of Electrical Engineers, *x*

Insulation:
  air-bubble impairment, 11, 33, 45, 112
  alcohol, 54
  bees wax, 48
  coils and condensers, 45, 47-49, 55
    Tesla patent, 12n
  gutta-percha, 49
  mica, 45, 46, 97
  oil, 12, 54
  paraffin, 48
  wax, 49, 105
  wax paper, 12

International Electrical Congress, *xi*

Interrupter, circuit; see Tesla coil, pri-- mary, break & circuit controller(s)

Inventions, Researches and Writings of Nikola Tesla, 3n, 4

Iridium, use of, 53, 104

Kelvin, Lord, 9, 20, 21, 36, 56

Laboratories; see Tesla, laboratories

Larsson, A., 22

Laue, Max von, 22

Lee Engineering Co., 4n

Lenard, Philipp, xv, *xvii*, 20, 31, 32, 100, 101, 112
  apparatus, 19, 23
  rays (or streams; see also cathodic streams, electron beams/rays), *xvii*, 1, 18-20, 22, 23, 29, 97, 100-102, 116
    magnetic deflection, 22, 100, 111, 114, 116
  vacuum tube (or bulb), *xvii*, 1, 2, 8, 19, 25, 26, 83, 86, 88, 94, 95, 97, 98, 102, 106

window, xvii, 20, 23

Leyden jars, 16

Light production efficiency, 11, 44

Lightning discharges, 102

Lodge, Oliver, *ix*

Longitudinal waves, 92

Magnetic circuit, 50

Marconi, Marchese Guglielmo, *ix*, 6

Mark Twain, 8, 8n

Martin, Thomas C., 3n, 4, 4n, 5, 6, 8n

Material stream emanations, 5, 21, 23-25, 87, 91-94, 96, 98, 100, 111, 116

Mauborgne, Joseph O., 16n

Meadowcroft, W.H., 115

Measurement methods:
angular velocity, 69, 116
curve-tracers, 66
electrostatic voltmeter, 47
frequency (or period), 2, 9n, 16n, 29, 33, 58, 65, 71
galvanometer, 114
motion, 17
oscilloscopes, 16
phase, 63
resistance, 66
stroboscope, electronic flash, 17, 58
mechanical, 17
Tesla instrument, 58-69, 71
voltage, 47
wave meters, 16, 16n

Medium (air, ether, natural), 18n, 25, 89, 100

Mercury, use of, *xiv*, 13, 14, 18, 19, 53, 74

Michelson-Morley experiment, *xiii*

Monell, S.H., 9n

Morton, William J., *xvi*

National Electric Light Association (NELA), *x*, 6

Negative glow region, 19

Neon flash lamp, 17
biasing method, 17

Neon signs, 19

Nernst lamp, 11

New York Academy of Sciences, *x*, *xvi*, *xvii*, 1, 1n, 2, 5, 12, 20, 22, 27, 76, 95, 100, 111, 112, 115

New York Electrical Society, *xi*

Newtonian physics, *xiii*

Nikola Tesla On His Work With Alternating Currents (1916 interview), *xvii*, 1, 1n, 2, 71

Oersted, Hans Christian, *xiv*

Oscillators, 5, 6n, 8, 11, 16, 34, 35, 39-41, 53, 71, 74, 74n, 75, (and "generators") 111

Oscillatory discharge(s), 11, 34, 36, 56, 68

Osterberg, Max, 115

Oxides, rare, 11

Oxygen, 50

Ozone, 49, 51, 84, 85, 93

Page, Charles Grafton, *xiv*

Particle-beam; see Material stream emanations
weaponry, *xvii*, 25, 25n

Perrin, Jean, *xiv*

Phase, adjustment of, 40

Phosphorescent coating; see X-ray, apparatus, screen

Piccard, Jean, *xiv*

Plate, photographic, 31, 32, 86, 91

Platinum:
hurtful actions associated with, 88, 90
use of, 85-87, 90, 92, 98

Plücker, Julius, *xiv*

Poggendorff, Johann, 12

Poggendorff, Johann, break investigations, 13, 13n, 53

Pohl, Ernst, 22

Positive column, 19

"powerful rays" (see also Ultraviolet rays), 113, 116

Practical Uses of the Wave Meter in Wireless Telegraphy, 16n

Pupin, Michael, 115

Radiant matter, *xiii*, *xv*

Radio:
patent, basic, *ix*
principles, first presented, *x*

Radioactivity, 21

Reis air thermometer, 73

Rhumkorff coil, *xiv*

Roentgen, Wilhelm Conrad, *x*, *xiii*, *xv*, *xvii*, 7, 8, 20, 25, 29, 32, 84, 95, 96, 99, 100, 112
apparatus, 19, 75
Nobel Prize, physics, 7

rays (or streams; see also X rays), *xvii*, 1, 1n, 2, 5n, 7, 7n, 16n, 18, 19, 21, 22, 29, 75, 83, 95, 97, 101, 102, 115, 116
nature, 100
source, 20, 21, 96, 97, 99, 115
screen, 98
tube (or bulb), 2, 8, 19, 26, 40, 42, 83, 88, 89

Roentgen Rays—and the Phenomena of the Anode and Cathode, 16n

Royal Institution, *x*

Rutherford, Ernest, 20

Savin, Marion; see Abrahams, Harold

Scientific American, 3

Secondary radiation; see X rays

Seifer, Marc, 4n

Sever, G.F., 115

Shallenberger, Oliver, 15
development of electric meter, 15
report of Tesla X-ray apparatus, 16

Siegrahn, M., 22

Single-electrode bulb(s), lamp(s), and tube(s), 1, 3, 10, 11, 16, 23, 24, 24n, 25, 31, 41, 75-81, 84-88, 91, 92, 94-102, 106
internal processes, 23, 24, 24n, 25
temperature limited, 25

Single-electrode tube(s):
directed discharges, 24n
particle-beam weapon, 25, 25n

Société Internationale des Electriciens, *x*

Société Francaise de Physique, *x*

Spark length; see Tesla coil, discharges

Standing waves; see Tesla coil, secondary

Stone, John Stone, *ix*

Story of Electricity, 6

Synchronization measurement, 18

Target; see Impact, body/plate

Telemechanics (wireless-controlled craft—*telautomata*), 3

Tesla coil(s), *xvii*, *xviii*, 8-10, 13, 16
desk-top units, 9, 10, 14, 15, 39 (Fig. 8), 41 (Fig. 9), 42, 43, 74
differences from induction coil(s), 33, 40, 42-44, 60
operation from municipal circuits, 9, 38, 40, 43, 45, 54, 75, 113
"Q," 15
secondary/primary turns ratio, 15
voltage gain, 15

Tesla coil:
discharges (sparks), 13, 43, 44, 51, 54-56
fluid-mechanics analogue, 8, 8n, 35, 57
original patent, 8n
primary,
augmenting distributed capacity in, 40
break (gap), 8, 16
dielectric, 34, 36, 54
circuit controller(s), 9, 39, 42, 54, 57, 58, 60, 66, 68, 69, 74, 75
patents, 9n, 74
condenser, 8
half-wavelength, 13-15, 56, 57
inductance, adjustment of, 58
interrupter, 40, 42, 43, 53, 54
dual-circuit, 39, 40
pseudo Thomson effect in, 13, 13n, 33, 56, 112
winding, 11
standing wave in, 15
circuit length, 13-15, 33, 55-57, 112
secondary,
standing wave in, 15
winding, 12
fine wire, objections to, 44
flat spiral, 15, 49
quarter-wavelength, 12, 15, 50
turns spacing, 49

Tesla experimental station, Colorado Springs, 13, 25

Tesla induction motor and AC systems patents, 16

Tesla laboratories:
35 South Fifth Ave., N.Y., 8, 8n, 10, 13, 31
destruction by fire, 7, 32, 112
46 East Houston St., N.Y., 10, 13, 15, 51, 71, 72

Tesla phosphor-coated lamps (see also Fluorescent lamps), 2, 10, 18, 31, 101
artificial daylight, 10, 11
first demonstration, 10
patent, 10
interference, 10

Tesla Trilogy, 4n

Tesla, Nikola:
arrival in America, *ix*
awards, *x*
Gregorian calendar birth date, *ix*
inventive gifts, *ix*
lectures, *x*
missed discovery of X rays, 7, 8, 32, 112, 116

Theoria Philosophiæ Naturalis, 21n

Therapeutic action, 84

Thompson, Edward P., 7n, 16n

Thomson, Elihu, 4, 10, 26, 27
  alleged controversy with Tesla, 27
  correspondence, 4n, 27
  patent interference with Tesla on single-electrode lamp, 10
  published exchanges with Tesla, 27
  X-ray burns, 27

Thomson-Houston Electric Co., 26

Thomson, Joseph, 20

Tonnelé & Co., 31, 112, 116

Trifunović, Dragan, 8n

Trio series lectures, 3, 10

Tuning, 9

Ultraviolet rays, 20

Units in text, 9n

Vacuum, 13, 18-20, 24, 40, 41, 47, 53-55, 104, 105
  bulb(s)/tube(s), xvii, xviii, 1, 2, 31, 44, 65, 68, 71, 75-81, 112, 115
  rotation of bodies in, 116

Vacuum-tube lighting; see Fluorescent/phosphorescent lamps

Visual synchronism; see Measurement methods, stroboscope, electronic-flash, Tesla instrument

Volta; Alessandro, *xiv* (see also Activity series of metals)

Waller, T., 22

Walter, Bernhard, 22

Western Electrician, 3

Westinghouse Electric & Manufacturing Co., 15, 26

Window, *xv*, *xvii*, 20, 23, 26, 38, 86-88, 92, 96-99, 102, 104, 106, 107

Wireless telegraphy (radio communication), *xvii*, 1, 1n, 3, 6, 15, 16n, 75
  patents, 3, 15, 18n
  receiving methods, 2, 18, 71-73

Wireless telephony, 1n
  patents, 3
  receiving methods, 72-74

Wireless transmission of power, patents, 1n, 3

Woodbury, David, 27, 27n

Working Manual of High Frequency Currents, 9n

Würzburg Physico-Medical Society, 7

WWI, 16

WWII, 25

X rays, 1n, 7, 16, 18-22, 22n, 23, 27
  hard, 19, 22
  mass absorption coefficient, 22
  medical application, *xvi*
  secondary radiation, 22
  soft, 19, 20, 22

X-ray:
  apparatus, 10, 19, 40, 41, 42, 101
  diffraction, 22
  directed discharges, 24
  discovery, *xiii*, *xv*, *xvii*, 2, 7, 29, 32, 84, 95, 100
  discovery announcement, 7, 8, 29, 32, 95, 96, 116
  fluorescent/phosphorescent screen, 29, 84, 85, 87, 91, 95, 98
  harms, 2, 26, 27, 83-95
    prevention, 26, 83, 85, 86, 92, 94, 106, 107
  methods, 1
  photography, 7n, 8, 31, 100, 112
  radiation, 5, 22, 27
  reflection (see also X rays, secondary radiation), *xvii*, *xviii*, 5n, 21, 22
  refraction, 22
  shadow graphs, distance, 23
  spectra, 22n
  tubes, 9, 21
    construction, 102-106
    double-focus, 103, 104
    potential limited by conducting objects, 23

Young, Thomas, *xiv*